BYRNE'S EUCLID

———

THE FIRST SIX BOOKS OF
THE ELEMENTS OF EUCLID
WITH COLOURED DIAGRAMS

AND SYMBOLS

THE FIRST SIX BOOKS OF

THE ELEMENTS OF EUCLID

IN WHICH COLOURED DIAGRAMS AND SYMBOLS

ARE USED INSTEAD OF LETTERS FOR THE

GREATER EASE OF LEARNERS

BY OLIVER BYRNE

SURVEYOR OF HER MAJESTY'S SETTLEMENTS IN THE FALKLAND ISLANDS
AND AUTHOR OF NUMEROUS MATHEMATICAL WORKS

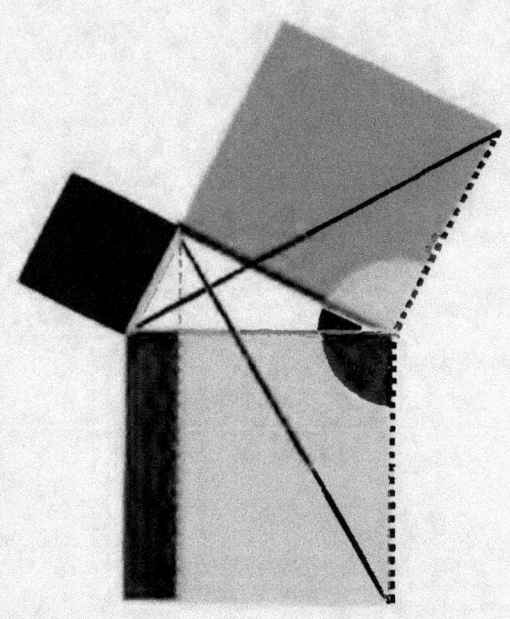

LONDON
WILLIAM PICKERING
1847

TO THE

RIGHT HONOURABLE THE EARL FITZWILLIAM,

ETC. ETC. ETC.

THIS WORK IS DEDICATED

BY HIS LORDSHIP'S OBEDIENT

AND MUCH OBLIGED SERVANT,

OLIVER BYRNE.

INTRODUCTION.

HE arts and sciences have become so extensive, that to facilitate their acquirement is of as much importance as to extend their boundaries. Illustration, if it does not shorten the time of study, will at least make it more agreeable. This Work has a greater aim than mere illustration; we do not introduce colours for the purpose of entertainment, or to amuse *by certain combinations of tint and form*, but to assist the mind in its researches after truth, to increase the facilities of instruction, and to diffuse permanent knowledge. If we wanted authorities to prove the importance and usefulness of geometry, we might quote every philosopher since the days of Plato. Among the Greeks, in ancient, as in the school of Pestalozzi and others in recent times, geometry was adopted as the best gymnastic of the mind. In fact, Euclid's Elements have become, by common consent, the basis of mathematical science all over the civilized globe. But this will not appear extraordinary, if we consider that this sublime science is not only better calculated than any other to call forth the spirit of inquiry, to elevate the mind, and to strengthen the reasoning faculties, but also it forms the best introduction to most of the useful and important vocations of human life. Arithmetic, land-surveying, mensuration, engineering, navigation, mechanics, hydrostatics, pneumatics, optics, physical astronomy, &c. are all dependent on the propositions of geometry.

Much however depends on the firſt communication of any ſcience to a learner, though the beſt and moſt eaſy methods are ſeldom adopted. Propoſitions are placed before a ſtudent, who though having a ſufficient underſtanding, is told juſt as much about them on entering at the very threſhold of the ſcience, as gives him a prepoſſeſſion moſt unfavourable to his future ſtudy of this delightful ſubject ; or " the formalities and paraphernalia of rigour are ſo oſtentatiouſly put forward, as almoſt to hide the reality. Endleſs and perplexing repetitions, which do not confer greater exactitude on the reaſoning, render the demonſtrations involved and obſcure, and conceal from the view of the ſtudent the conſecution of evidence." Thus an averſion is created in the mind of the pupil, and a ſubject ſo calculated to improve the reaſoning powers, and give the habit of cloſe thinking, is degraded by a dry and rigid courſe of inſtruction into an unintereſting exerciſe of the memory. To raiſe the curioſity, and to awaken the liſtleſs and dormant powers of younger minds ſhould be the aim of every teacher ; but where examples of excellence are wanting, the attempts to attain it are but few, while eminence excites attention and produces imitation. The object of this Work is to introduce a method of teaching geometry, which has been much approved of by many ſcientific men in this country, as well as in France and America. The plan here adopted forcibly appeals to the eye, the moſt ſenſitive and the moſt comprehenſive of our external organs, and its pre-eminence to imprint it ſubject on the mind is ſupported by the incontrovertible maxim expreſſed in the well known words of Horace :—

Segnius irritant animos demiſſa per aurem
Quàm quæ ſunt oculis ſubjecta fidelibus.

A feebler impreſs through the ear is made,
Than what is by the faithful eye conveyed.

All language confifts of reprefentative figns, and thofe figns are the beft which effect their purpofes with the greateft precifion and difpatch. Such for all common purpofes are the audible figns called words, which are ftill confidered as audible, whether addreffed immediately to the ear, or through the medium of letters to the eye. Geometrical diagrams are not figns, but the materials of geometrical fcience, the object of which is to fhow the relative quantities of their parts by a procefs of reafoning called Demonftration. This reafoning has been generally carried on by words, letters, and black or uncoloured diagrams; but as the ufe of coloured fymbols, figns, and diagrams in the linear arts and fciences, renders the procefs of reafoning more precife, and the attainment more expeditious, they have been in this inftance accordingly adopted.

Such is the expedition of this enticing mode of communicating knowledge, that the Elements of Euclid can be acquired in lefs than one third the time ufually employed, and the retention by the memory is much more permanent; thefe facts have been afcertained by numerous experiments made by the inventor, and feveral others who have adopted his plans. The particulars of which are few and obvious; the letters annexed to points, lines, or other parts of a diagram are in fact but arbitrary names, and reprefent them in the demonftration; inftead of thefe, the parts being differently coloured, are made to name themfelves, for their forms in correfponding colours represent them in the demonftration.

In order to give a better idea of this fyftem, and of the advantages gained by its adoption, let us take a right angled triangle, and expreſs ſome of its properties both by colours and the method generally employed.

Some of the properties of the right angled triangle ABC, expreſſed by the method generally employed.

1. The angle BAC, together with the angles BCA and ABC are equal to two right angles, or twice the angle ABC.

2. The angle CAB added to the angle ACB will be equal to the angle ABC.

3. The angle ABC is greater than either of the angles BAC or BCA.

4. The angle BCA or the angle CAB is leſs than the angle ABC.

5. If from the angle ABC, there be taken the angle BAC, the remainder will be equal to the angle ACB.

6. The ſquare of AC is equal to the ſum of the ſquares of AB and BC.

The ſame properties expreſſed by colouring the different parts.

I. 

That is, the red angle added to the yellow angle added to the blue angle, equal twice the yellow angle, equal two right angles.

2. 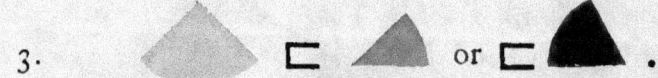

Or in words, the red angle added to the blue angle, equal the yellow angle.

3.

The yellow angle is greater than either the red or blue angle.

4.

Either the red or blue angle is lefs than the yellow angle.

5.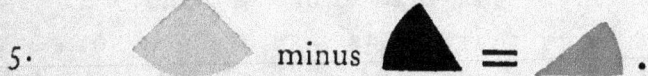

In other terms, the yellow angle made lefs by the blue angle equal the red angle.

6. $$\rule{2cm}{0.4pt}{}^{2} = \rule{2cm}{1pt}{}^{2} + \rule{2cm}{0.4pt}{}^{2}.$$

That is, the fquare of the yellow line is equal to the fum of the fquares of the blue and red lines.

In oral demonftrations we gain with colours this important advantage, the eye and the ear can be addreffed at the fame moment, fo that for teaching geometry, and other linear arts and fciences, in claffes, the fyftem is the beft ever propofed, this is apparent from the examples juft given.

Whence it is evident that a reference from the text to the diagram is more rapid and fure, by giving the forms and colours of the parts, or by naming the parts and their colours, than naming the parts and letters on the diagram. Befides the fuperior fimplicity, this fyftem is likewife confpicuous for concentration, and wholly excludes the injurious though prevalent practice of allowing the ftudent to commit the demonftration to memory; until reafon, and fact, and proof only make impreffions on the underftanding.

Again, when lecturing on the principles or properties of figures, if we mention the colour of the part or parts referred to, as in faying, the red angle, the blue line, or lines, &c. the part or parts thus named will be immediately feen by all in the clafs at the fame inftant; not fo if we fay the angle ABC, the triangle PFQ, the figure EGKt, and fo on;

for the letters muſt be traced one by one before the ſtudents arrange in their minds the particular magnitude referred to, which often occaſions confuſion and error, as well as loſs of time. Alſo if the parts which are given as equal, have the ſame colours in any diagram, the mind will not wander from the object before it; that is, ſuch an arrangement preſents an ocular demonſtration of the parts to be proved equal, and the learner retains the data throughout the whole of the reaſoning. But whatever may be the advantages of the preſent plan, if it be not ſubſtituted for, it can always be made a powerful auxiliary to the other methods, for the purpoſe of introduction, or of a more ſpeedy reminiſcence, or of more permanent retention by the memory.

The experience of all who have formed ſyſtems to impreſs facts on the underſtanding, agree in proving that coloured repreſentations, as pictures, cuts, diagrams, &c. are more eaſily fixed in the mind than mere ſentences unmarked by any peculiarity. Curious as it may appear, poets ſeem to be aware of this fact more than mathematicians; many modern poets allude to this viſible ſyſtem of communicating knowledge, one of them has thus expreſſed himſelf:

> Sounds which addreſs the ear are loſt and die
> In one ſhort hour, but theſe which ſtrike the eye,
> Live long upon the mind, the faithful ſight
> Engraves the knowledge with a beam of light.

This perhaps may be reckoned the only improvement which plain geometry has received ſince the days of Euclid, and if there were any geometers of note before that time, Euclid's ſucceſs has quite eclipſed their memory, and even occaſioned all good things of that kind to be aſſigned to him; like Æſop among the writers of Fables. It may alſo be worthy of remark, as tangible diagrams afford the only medium through which geometry and other linear arts and ſciences can be taught to the blind, this viſible ſyſtem is no leſs adapted to the exigencies of the deaf and dumb.

Care muſt be taken to ſhow that colour has nothing to do with the lines, angles, or magnitudes, except merely to name them. A mathematical line, which is length without breadth, cannot poſſeſs colour, yet the junction of two colours on the ſame plane gives a good idea of what is meant by a mathematical line; recollect we are ſpeaking familiarly, ſuch a junction is to be underſtood and not the colour, when we ſay the black line, the red line or lines, &c.

Colours and coloured diagrams may at firſt appear a clumſy method to convey proper notions of the properties and parts of mathematical figures and magnitudes, however they will be found to afford a means more refined and extenſive than any that has been hitherto propoſed.

We ſhall here define a point, a line, and a ſurface, and demonſtrate a propoſition in order to ſhow the truth of this aſſertion.

A point is that which has poſition, but not magnitude; or a point is poſition only, abſtracted from the conſideration of length, breadth, and thickneſs. Perhaps the following deſcription is better calculated to explain the nature of a mathematical point to thoſe who have not acquired the idea, than the above ſpecious definition.

Let three colours meet and cover a portion of the paper, where they meet is not blue, nor is it yellow, nor is it red, as it occupies no portion of the plane, for if it did, it would belong to the blue, the red, or the yellow part; yet it exiſts, and has poſition without magnitude, ſo that with a little reflection, this junc-

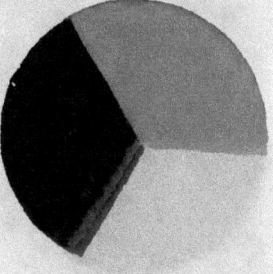

tion of three colours on a plane, gives a good idea of a mathematical point.

A line is length without breadth. With the affiftance of colours, nearly in the fame manner as before, an idea of a line may be thus given :—

Let two colours meet and cover a portion of the paper;

where they meet is not red, nor is it blue; therefore the junction occupies no portion of the plane, and therefore it cannot have breadth, but only length : from which we can readily form an idea of what is meant by a mathematical line. For the purpofe of illuftration, one colour differing from the colour of the paper, or plane upon which it is drawn, would have been fufficient; hence in future, if we fay the red line, the blue line, or lines, &c. it is the junctions with the plane upon which they are drawn are to be underftood.

Surface is that which has length and breadth without thicknefs.

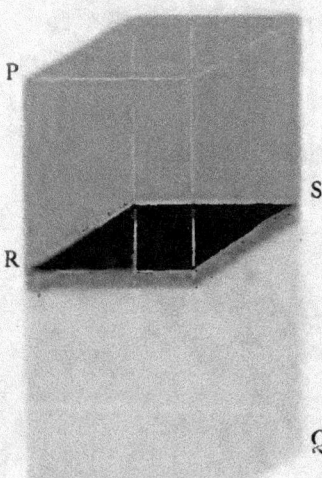

When we confider a folid body (**PQ**), we perceive at once that it has three dimenfions, namely :— length, breadth, and thicknefs; fuppofe one part of this folid (PS) to be red, and the other part (**QR**) yellow, and that the colours be diftinct without commingling, the blue furface (RS) which feparates thefe parts, or which is the fame thing, that which divides the folid without lofs of material, muft be without thicknefs, and only poffeffes length and breadth;

INTRODUCTION.   XV

this plainly appears from reasoning, similar to that just employed in defining, or rather describing a point and a line.

The proposition which we have selected to elucidate the manner in which the principles are applied, is the fifth of the first Book.

In an isosceles triangle ABC, the internal angles at the base ABC, ACB are equal, and when the sides AB, AC are produced, the external angles at the base BCE, CBD are also equal.

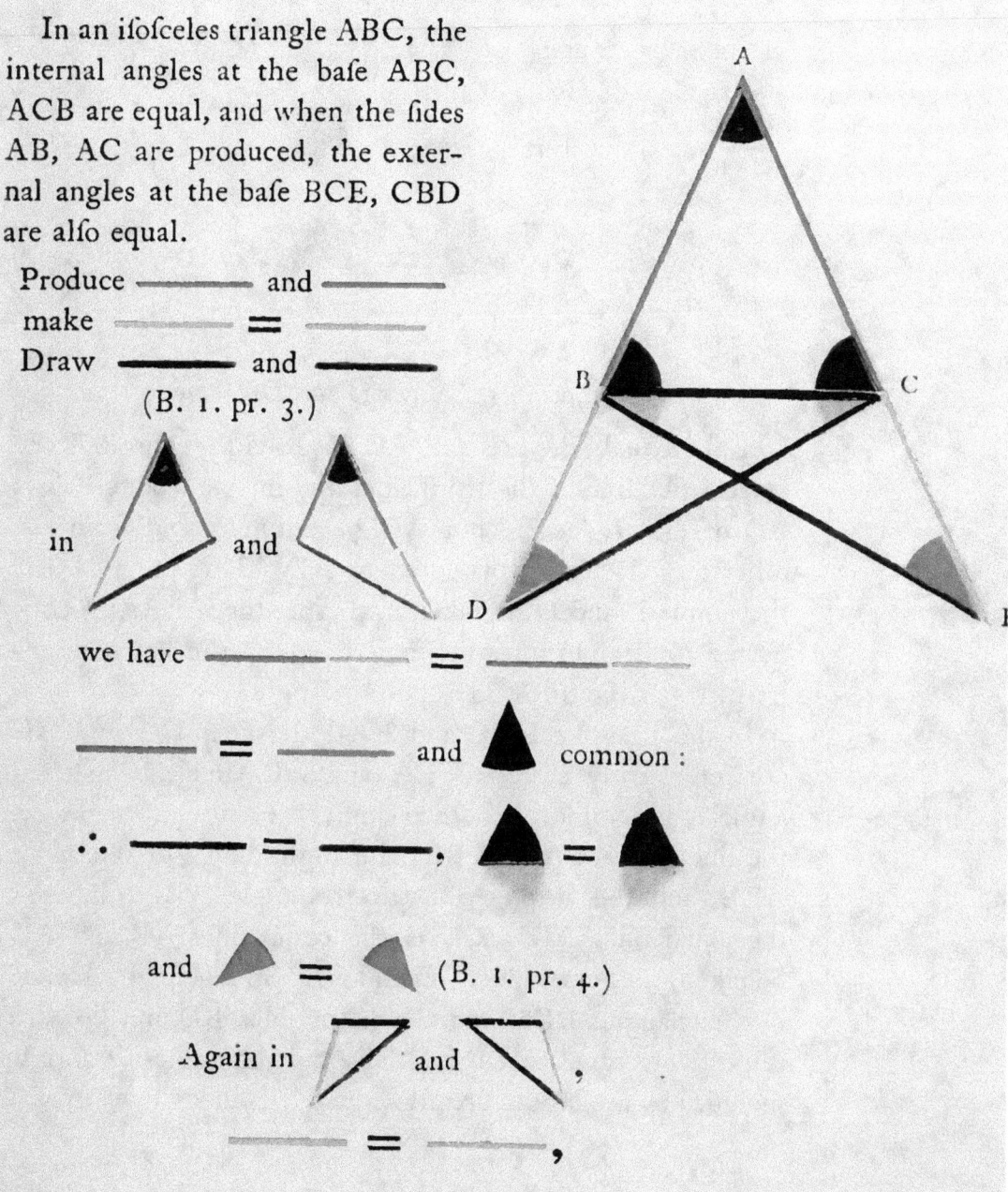

Produce ——— and ———

make ——— ═ ———

Draw ——— and ———

(B. 1. pr. 3.)

in ▲ and ▲

we have ——— ═ ———

——— ═ ——— and ▲ common :

∴ ——— ═ ———, ▲ ═ ▲

and ◣ ═ ◢ (B. 1. pr. 4.)

Again in ◺ and ◹,

——— ═ ———,

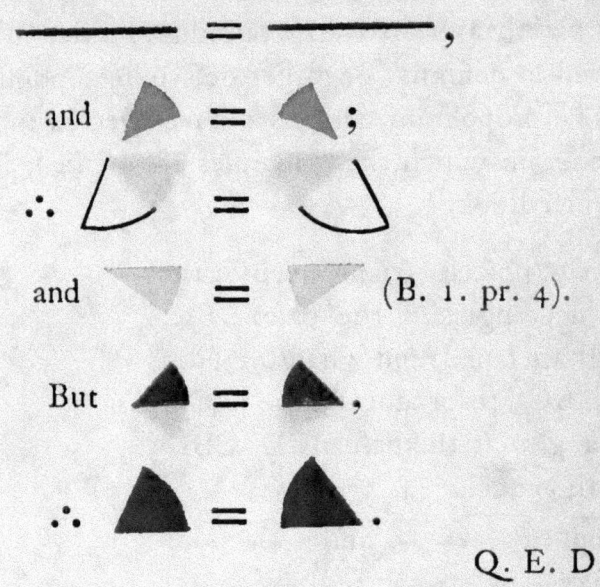

Q. E. D.

By annexing Letters to the Diagram.

Let the equal fides AB and AC be produced through the extremities BC, of the third fide, and in the produced part BD of either, let any point D be affumed, and from the other let AE be cut off equal to AD (B. 1. pr. 3). Let the points E and D, fo taken in the produced fides, be connected by ftraight lines DC and BE with the alternate extremities of the third fide of the triangle.

In the triangles DAC and EAB the fides DA and AC are refpectively equal to EA and AB, and the included angle A is common to both triangles. Hence (B. 1. pr. 4.) the line DC is equal to BE, the angle ADC to the angle AEB, and the angle ACD to the angle ABE; if from the equal lines AD and AE the equal fides AB and AC be taken, the remainders BD and CE will be equal. Hence in the triangles BDC and CEB, the fides BD and DC are refpectively equal to CE and EB, and the angles D and E included by thofe fides are alfo equal. Hence (B. 1. pr. 4.)

the angles DBC and ECB, which are thofe included by the third fide BC and the productions of the equal fides AB and AC are equal. Alfo the angles DCB and EBC are equal if thofe equals be taken from the angles DCA and EBA before proved equal, the remainders, which are the angles ABC and ACB oppofite to the equal fides, will be equal.

Therefore in an ifofceles triangle, &c.

Q. E. D.

Our object in this place being to introduce the fyftem rather than to teach any particular fet of propofitions, we have therefore felected the foregoing out of the regular courfe. For fchools and other public places of inftruction, dyed chalks will anfwer to defcribe diagrams, &c. for private ufe coloured pencils will be found very convenient.

We are happy to find that the Elements of Mathematics now forms a confiderable part of every found female education, therefore we call the attention of thofe interefted or engaged in the education of ladies to this very attractive mode of communicating knowledge, and to the fucceeding work for its future developement.

We fhall for the prefent conclude by obferving, as the fenfes of fight and hearing can be fo forcibly and inftantaneoufly addreffed alike with one thoufand as with one, *the million* might be taught geometry and other branches of mathematics with great eafe, this would advance the purpofe of education more than any thing that *might* be named, for it would teach the people how to think, and not what to think; it is in this particular the great error of education originates.

THE ELEMENTS OF EUCLID.

BOOK I.

DEFINITIONS.

I.

A *point* is that which has no parts.

II.

A *line* is length without breadth.

III.

The extremities of a line are points.

IV.

A ſtraight or right line is that which lies evenly between its extremities.

V.

A ſurface is that which has length and breadth only.

VI.

The extremities of a ſurface are lines.

VII.

A plane ſurface is that which lies evenly between its extremities.

VIII.

A plane angle is the inclination of two lines to one another, in a plane, which meet together, but are not in the ſame direction.

IX.

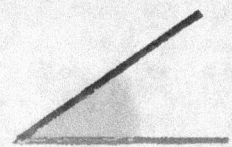
A plane rectilinear angle is the inclination of two ſtraight lines to one another, which meet together, but are not in the ſame ſtraight line.

X.

When one ftraight line ftanding on another ftraight line makes the adjacent angles
equal, each of thefe angles is called a *right
angle*, and each of thefe lines is faid to be
perpendicular to the other.

XI.

An obtufe angle is an angle greater
than a right angle.

XII.

An acute angle is an angle lefs than a
right angle.

XIII.

A term or boundary is the extremity of any thing.

XIV.

A figure is a furface enclofed on all fides by a line or lines.

XV.

A circle is a plane figure, bounded
by one continued line, called its circumference or periphery; and having a certain point within it, from
which all ftraight lines drawn to its
circumference are equal.

XVI.

This point (from which the equal lines are drawn) is
called the centre of the circle.

XVII.

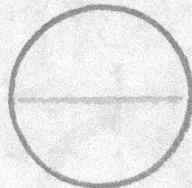

A diameter of a circle is a ſtraight line drawn through the centre, terminated both ways in the circumference.

XVIII.

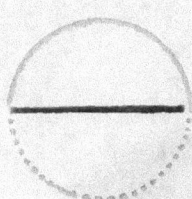

A ſemicircle is the figure contained by the diameter, and the part of the circle cut off by the diameter.

XIX.

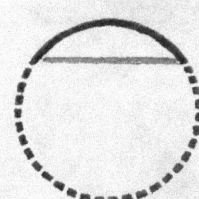

A ſegment of a circle is a figure contained by a ſtraight line, and the part of the circumference which it cuts off.

XX.

A figure contained by ſtraight lines only, is called a rectilinear figure.

XXI.

A triangle is a rectilinear figure included by three ſides.

XXII.

A quadrilateral figure is one which is bounded by four ſides. The ſtraight lines ▬▬▬ and ▬▬▬ connecting the vertices of the oppoſite angles of a quadrilateral figure, are called its diagonals.

XXIII.

A polygon is a rectilinear figure bounded by more than four ſides.

XXIV.

A triangle whofe three fides are equal, is faid to be equilateral.

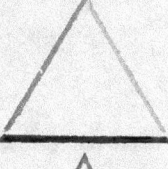

XXV.

A triangle which has only two fides equal is called an ifofceles triangle.

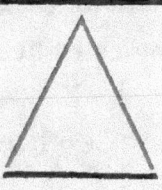

XXVI.

A fcalene triangle is one which has no two fides equal.

XXVII.

A right angled triangle is that which has a right angle.

XXVIII.

An obtufe angled triangle is that which has an obtufe angle.

XXIX.

An acute angled triangle is that which has three acute angles.

XXX.

Of four-fided figures, a fquare is that which has all its fides equal, and all its angles right angles.

XXXI.

A rhombus is that which has all its fides equal, but its angles are not right angles.

XXXII.

An oblong is that which has all its angles right angles, but has not all its fides equal.

XXXIII.

A rhomboid is that which has its opposite fides equal to one another, but all its fides are not equal, nor its angles right angles.

XXXIV.

All other quadrilateral figures are called trapeziums.

XXXV.

Parallel ftraight lines are fuch as are in the fame plane, and which being produced continually in both directions, would never meet.

POSTULATES.

I.

Let it be granted that a ftraight line may be drawn from any one point to any other point.

II.

Let it be granted that a finite ftraight line may be produced to any length in a ftraight line.

III.

Let it be granted that a circle may be defcribed with any centre at any diftance from that centre.

AXIOMS.

I.

Magnitudes which are equal to the fame are equal to each other.

II.

If equals be added to equals the fums will be equal.

III.

If equals be taken away from equals the remainders will be equal.

IV.

If equals be added to unequals the fums will be unequal.

V.

If equals be taken away from unequals the remainders will be unequal.

VI.

The doubles of the fame or equal magnitudes are equal.

VII.

The halves of the fame or equal magnitudes are equal.

VIII.

Magnitudes which coincide with one another, or exactly fill the fame fpace, are equal.

IX.

The whole is greater than its part.

X.

Two ftraight lines cannot include a fpace.

XI.

All right angles are equal.

XII.

If two ftraight lines (————) meet a third ftraight line (———) fo as to make the two interior angles (and) on the fame fide lefs than two right angles, thefe two ftraight lines will meet if they be produced on that fide on which the angles are lefs than two right angles.

The twelfth axiom may be expreſſed in any of the following ways :

1. Two diverging ſtraight lines cannot be both parallel
to the ſame ſtraight line.

2. If a ſtraight line interſect one of the two parallel
ſtraight lines it muſt alſo interſect the other.

3. Only one ſtraight line can be drawn through a given
point, parallel to a given ſtraight line.

Geometry has for its principal objects the expoſition and
explanation of the properties of *figure*, and figure is defined
to be the relation which ſubſiſts between the boundaries of
ſpace.　Space or magnitude is of three kinds, *linear*, *ſuper-
ficial*, and *ſolid*.

Angles might properly be conſidered as a fourth ſpecies
of magnitude.　Angular magnitude evidently conſiſts of
parts, and muſt therefore be admitted to be a ſpecies of
quantity　The ſtudent muſt not ſuppoſe that the magnitude of an angle is affected by the length
of the ſtraight lines which include it, and
of whoſe mutual divergence it is the mea-
ſure.　The *vertex* of an angle is the point
where the *ſides* or the *legs* of the angle
meet, as A.

An angle is often deſignated by a ſingle letter when its
legs are the only lines which meet to-
gether at its vertex. Thus the red and
blue lines form the yellow angle, which
in other ſyſtems would be called the
angle A.　But when more than two
lines meet in the ſame point, it was ne-
ceſſary by former methods, in order to
avoid confuſion, to employ three letters
to deſignate an angle about that point,

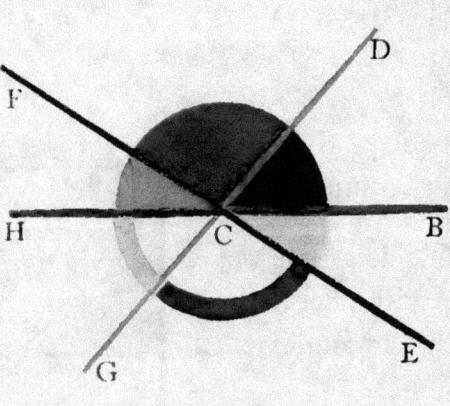

the letter which marked the vertex of the angle being always placed in the middle. Thus the black and red lines meeting together at C, form the blue angle, and has been ufually denominated the angle FCD or DCF The lines FC and CD are the legs of the angle; the point C is its vertex. In like manner the black angle would be defignated the angle DCB or BCD. The red and blue angles added together, or the angle HCF added to FCD, make the angle HCD; and fo of other angles.

When the legs of an angle are produced or prolonged beyond its vertex, the angles made by them on both fides of the vertex are faid to be *vertically oppofite* to each other: Thus the red and yellow angles are faid to be vertically oppofite angles.

Superpofition is the procefs by which one magnitude may be conceived to be placed upon another, fo as exactly to cover it, or fo that every part of each fhall exactly coincide.

A line is faid to be *produced*, when it is extended, prolonged, or has its length increafed, and the increafe of length which it receives is called its *produced part*, or its *production*.

The entire length of the line or lines which enclofe a figure, is called its *perimeter*. The firft fix books of Euclid treat of plain figures only. A line drawn from the centre of a circle to its circumference, is called a *radius*. The lines which include a figure are called its *fides*. That fide of a right angled triangle, which is oppofite to the right angle, is called the *hypotenufe*. An *oblong* is defined in the fecond book, and called a *rectangle*. All the lines which are confidered in the firft fix books of the Elements are fuppofed to be in the fame plane.

The *ftraight-edge* and *compaffes* are the only inftruments, the use of which is permitted in Euclid, or plain Geometry. To declare this reftriction is the object of the *poftulates*.

The *Axioms* of geometry are certain general propofitions, the truth of which is taken to be felf-evident and incapable of being eftablifhed by demonftration.

Propofitions are thofe refults which are obtained in geometry by a procefs of reafoning. There are two fpecies of propofitions in geometry, *problems* and *theorems*.

A *Problem* is a propofition in which fomething is propofed to be done; as a line to be drawn under fome given conditions, a circle to be defcribed, fome figure to be conftructed, &c.

The *folution* of the problem confifts in fhowing how the thing required may be done by the aid of the rule or ftraight-edge and compaffes.

The *demonftration* confifts in proving that the procefs indicated in the folution really attains the required end.

A *Theorem* is a propofition in which the truth of fome principle is afferted. This principle muft be deduced from the axioms and definitions, or other truths previoufly and independently eftablifhed. To fhow this is the object of demonftration.

A *Problem* is analogous to a poftulate.

A *Theorem* refembles an axiom.

A *Poftulate* is a problem, the folution of which is affumed.

An *Axiom* is a theorem, the truth of which is granted without demonftration.

A *Corollary* is an inference deduced immediately from a propofition.

A *Scholium* is a note or obfervation on a propofition not containing an inference of fufficient importance to entitle it to the name of a *corollary*.

A *Lemma* is a propofition merely introduced for the purpofe of eftablifhing fome more important propofition.

SYMBOLS AND ABBREVIATIONS.

∴ expreſſes the word *therefore*.

∵ . . . . *becauſe*.

= . . . . *equal*. This ſign of equality may be read *equal to*, or *is equal to*, or *are equal to ;* but any diſcrepancy in regard to the introduction of the auxiliary verbs *is*, *are*, &c. cannot affect the geometrical rigour.

≠ means the ſame as if the words ' *not equal*' were written.

⊐ ſignifies *greater than*.

⊏ . . . . *leſs than*.

⊯ . . . . *not greater than*.

⊮ . . . . *not leſs than*.

+ is read *plus (more)*, the ſign of addition ; when interpoſed between two or more magnitudes, ſignifies their ſum.

— is read *minus (leſs)*, ſignifies ſubtraction ; and when placed between two quantities denotes that the latter is to be taken from the former.

× this ſign expreſſes the product of two or more numbers when placed between them in arithmetic and algebra ; but in geometry it is generally uſed to expreſs a *rectangle*, when placed between " two ſtraight lines which contain one of its right angles." A *rectangle* may alſo be repreſented by placing a point between two of its conterminous ſides.

: :: : expreſſes an *analogy* or *proportion ;* thus, if A, B, C and D, repreſent four magnitudes, and A has to B the ſame ratio that C has to D, the propoſition is thus briefly written,

$$A : B :: C : D,$$
$$A : B = C : D,$$
$$\text{or } \frac{A}{B} = \frac{C}{D}.$$

This equality or ſameneſs of ratio is read,

as A is to B, fo is C to D;

or A is to B, as C is to D.

$\parallel$ fignifies *parallel to.*

$\perp$ . . . . *perpendicular to.*

$\triangle$. *angle.*

. . *right angle.*

two right angles.

or $\wedge$ briefly defignates a *point*.

$\sqsubset$, $=$, or $\sqsupset$ fignifies *greater, equal, or lefs than.*

The fquare defcribed on a line is concifely written thus,
_____².

In the fame manner twice the fquare of, is expreffed by
2 . _____².

def. fignifies *definition.*

pos. . . . . *poftulate.*

ax. . . . . *axiom.*

hyp. . . . . *hypothefis.* It may be neceffary here to remark, that the *hypothefis* is the condition affumed or
taken for granted. Thus, the hypothefis of the propofition given in the Introduction, is that the triangle
is ifofceles, or that its legs are equal.

conft. . . . . *conftruction.* The *conftruction* is the change
made in the original figure, by drawing lines, making
angles, defcribing circles, &c. in order to adapt it to
the argument of the demonftration or the folution of
the problem. The conditions under which thefe
changes are made, are as indifputable as thofe contained in the hypothefis. For inftance, if we make
an angle equal to a given angle, thefe two angles are
equal by conftruction.

Q. E. D. . . . . *Quod erat demonftrandum.*

Which was to be demonftrated.

Faults to be corrected before reading this Volume.

PAGE 13, line 9, *for* def. 7 *read* def. 10.

    45, laft line, *for* pr. 19 *read* pr. 29.

    54, line 4 from the bottom, *for* black and red line *read* blue and red line.

    59, line 4, *for* add black line fquared *read* add blue line fquared.

    60, line 17, *for* red line multiplied by red and yellow line *read* red line multiplied by red, blue, and yellow line.

    76, line 11, *for* def. 7 *read* def. 10.

    81, line 10, *for* take black line *read* take blue line.

    105, line 11, *for* yellow black angle add blue angle equal red angle *read* yellow black angle add blue angle add red angle.

    129, laft line, *for* circle *read* triangle.

    141, line 1, *for* Draw black line *read* Draw blue line.

    196, line 3, before the yellow magnitude infert M.

Euclid.

BOOK I.

PROPOSITION I. PROBLEM.

ON a given finite *straight line* (——) *to describe an equila-teral triangle.*

Describe  and

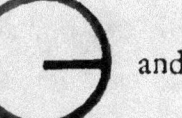 (poſtulate 3.); draw ——— and —— (poſt. 1.).

then will ▲ be equilateral.

For —— = ——— (def. 15.);

and —— = —— (def. 15.),

∴ ——— = —— (axiom. 1.);

and therefore ▲ is the equilateral triangle required.

Q. E. D.

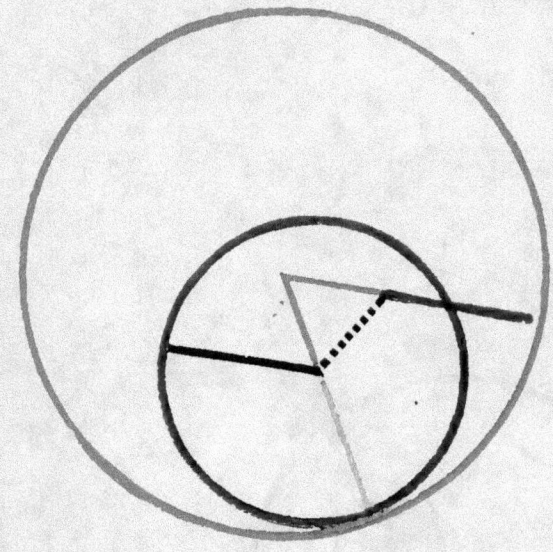

ROM *a given point* (———），
*to draw a straight line equal
to a given finite straight
line* (———).

Draw ·········· (post. 1.), describe

(pr. 1.), produce ——— (post.

2.), describe  (post. 3.), and

(post. 3.); produce ——— (post. 2.), then

——— is the line required.

For ——— = ——— (def. 15.),

and ——— = ——— (const.), ∴ ——— = ———

(ax. 3.), but (def. 15.) ——— = ——— = ———;

∴ ——— drawn from the given point (———),

is equal the given line ———.

Q. E. D.

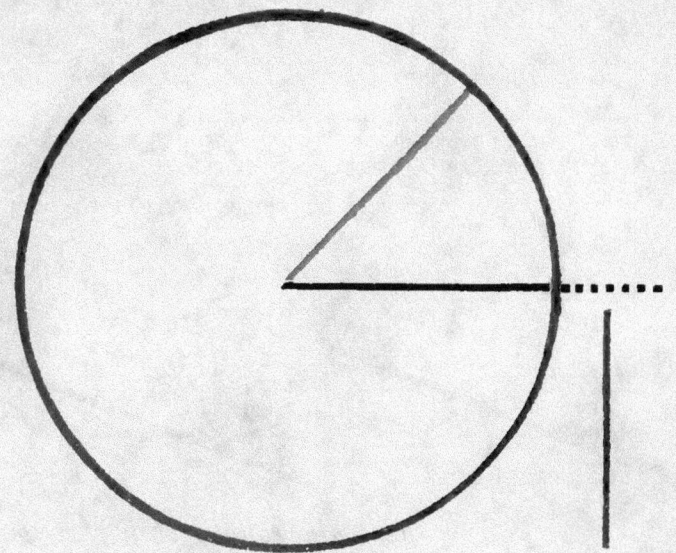

ROM *the greater* (———......) *of two given straight lines, to cut off a part equal to the less* (———).

Draw 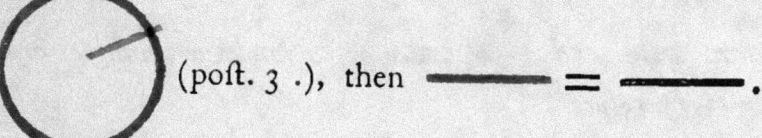 (pr. 2.); defcribe (poft. 3 .), then ——— = ———.

For ——— = ——— (def. 15.),

and ——— = ——— (conft.);

∴ ——— = ——— (ax. 1.).

Q. E. D.

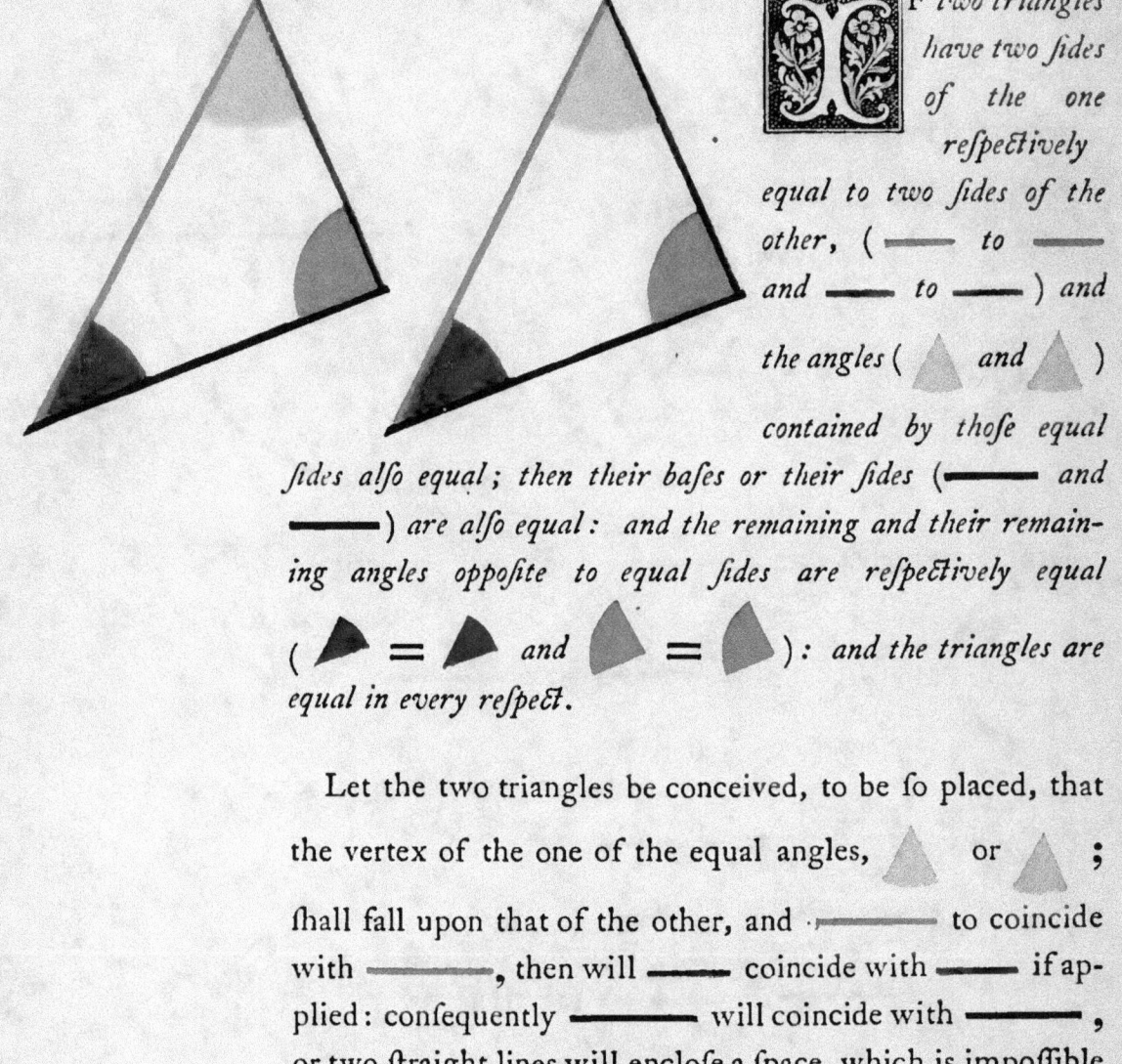

F *two triangles have two sides of the one respectively equal to two sides of the other, (———— to ———— and ———— to ————) and the angles (▲ and ▲) contained by those equal sides also equal; then their bases or their sides (———— and ————) are also equal: and the remaining and their remaining angles opposite to equal sides are respectively equal (◢ = ◣ and ◗ = ◗): and the triangles are equal in every respect.*

Let the two triangles be conceived, to be so placed, that the vertex of the one of the equal angles, ▲ or ▲ ; shall fall upon that of the other, and ———— to coincide with ————, then will ———— coincide with ———— if applied: consequently ———— will coincide with ————, or two straight lines will enclose a space, which is impossible (ax. 10), therefore ———— = ————, ◢ = ◣

and ◗ = ◗, and as the triangles △ and △ coincide, when applied, they are equal in every respect.

Q. E. D.

 N *any isosceles triangle* 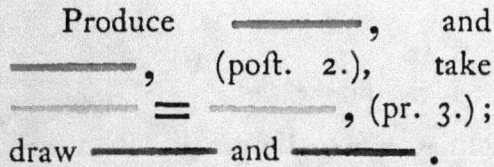 *if the equal sides be produced, the external angles at the base are equal, and the internal angles at the base are also equal.*

Produce ———, and ———, (post. 2.), take ——— = ———, (pr. 3.); draw ——— and ———.

Then in 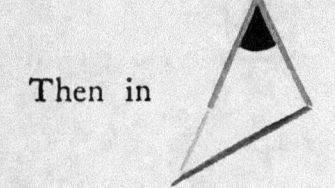 and  we have,

——— = ——— (const.), ▲ common to both, and ——— = ——— (hyp.) ∴ ◣ = ◢,

——— = ——— and ◣ = ◢ (pr. 4.).

Again in ◺◹ and we have ——— = ———,

◣ = ◢ and ——— = ———, ∴

◡ = ◡ and ◺ = ◹ (pr. 4.) but

◣ = ◢, ∴ ◣ = ◢ (ax. 3.)

Q. E. D.

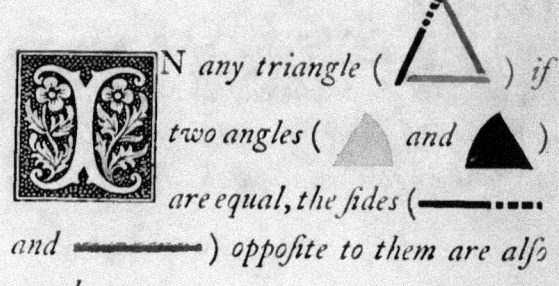 N *any triangle* ( ) *if two angles* (and) *are equal, the fides* (——— ·····) *and* (———) *oppofite to them are alfo equal.*

For if the fides be not equal, let one of them ——— ····· be greater than the other ———, and from it cut off ——— = ——— (pr. 3.), draw ··········· .

Then in and , ——— ····· = ———, (conft.) = (hyp.) and ——— common, ∴ the triangles are equal (pr. 4.) a part equal to the whole, which is abfurd; ∴ neither of the fides ——— ····· or ——— is greater than the other, ∴ hence they are equal

Q. E. D.

O N *the fame bafe (* ⸺ *), and on the fame fide of it there cannot be two triangles having their conterminous fides (* ⸺ *and* ⸺ *,* ⸺ *and* ⸺ *) at both extremities of the bafe, equal to each other.*

When two triangles ftand on the fame bafe, and on the fame fide of it, the vertex of the one fhall either fall outfide of the other triangle, or within it; or, laftly, on one of its fides.

If it be poffible let the two triangles be con-ftructed fo that $\left\{\begin{array}{c} \rule{3em}{0.4pt} = \rule{3em}{0.4pt} \\ \rule{3em}{0.4pt} = \rule{3em}{0.4pt} \end{array}\right\}$, then

draw ▬ ▬ ▬ and,

◨ = ▼ (pr. 5.)

∴ ▼ ⊐ ▼ and

∴ ▼ ⊐ ◤ $\left.\begin{array}{c} \\ \\ \\ \end{array}\right\}$ which is abfurd,

but (pr. 5.) ▼ = ◤

therefore the two triangles cannot have their conterminous fides equal at both extremities of the bafe.

Q. E. D.

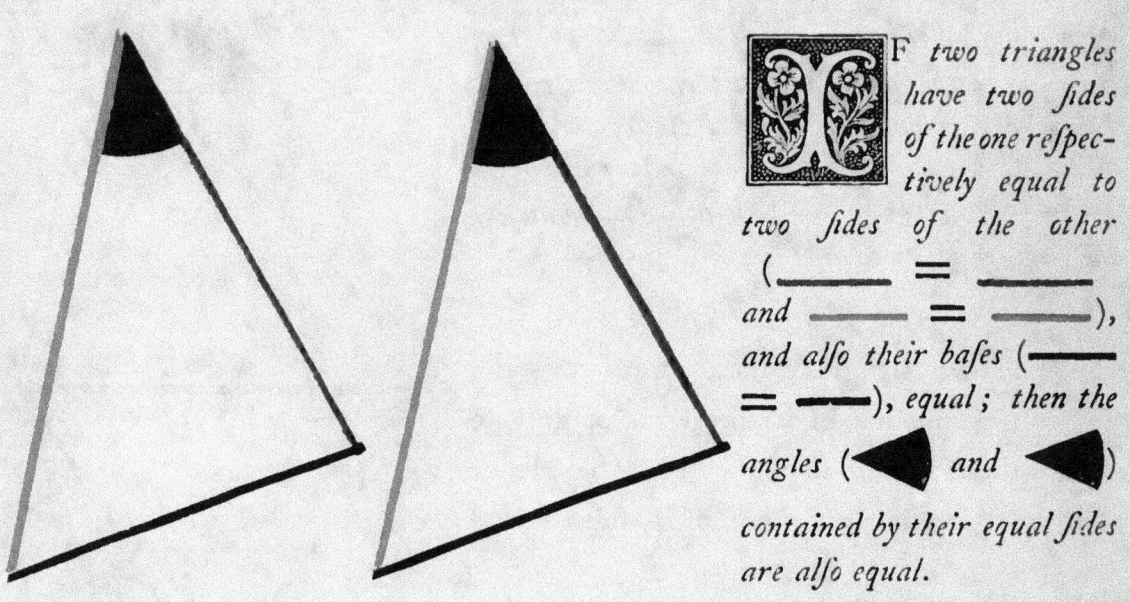

F *two triangles have two sides of the one respectively equal to two sides of the other* (_____ = _____ *and* _____ = _____), *and also their bases* (_____ = _____), *equal; then the angles* (◀ *and* ◀) *contained by their equal sides are also equal.*

If the equal bases _____ and _____ be conceived to be placed one upon the other, so that the triangles shall lie at the same side of them, and that the equal sides _____ and _____, _____ and _____ be conterminous, the vertex of the one must fall on the vertex of the other; for to suppose them not coincident would contradict the last proposition.

Therefore the sides _____ and _____, being coincident with _____ and _____,

∴ ▲ = ▲ .

Q. E. D.

O *bifect a given rectilinear* *angle* (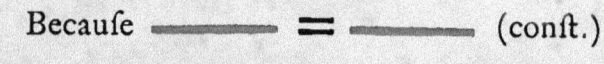).

Take ▬▬ = ▬▬ (pr. 3.)
draw ▬▬ , upon which

defcribe ⋁ (pr. 1.),
draw ▬▬ .

Becaufe ▬▬ = ▬▬ (conft.)
and ▬▬ common to the two triangles
and ▬▬ = ▬▬ (conft.),

∴ ◀ = ◣ (pr. 8.)

Q. E. D.

c

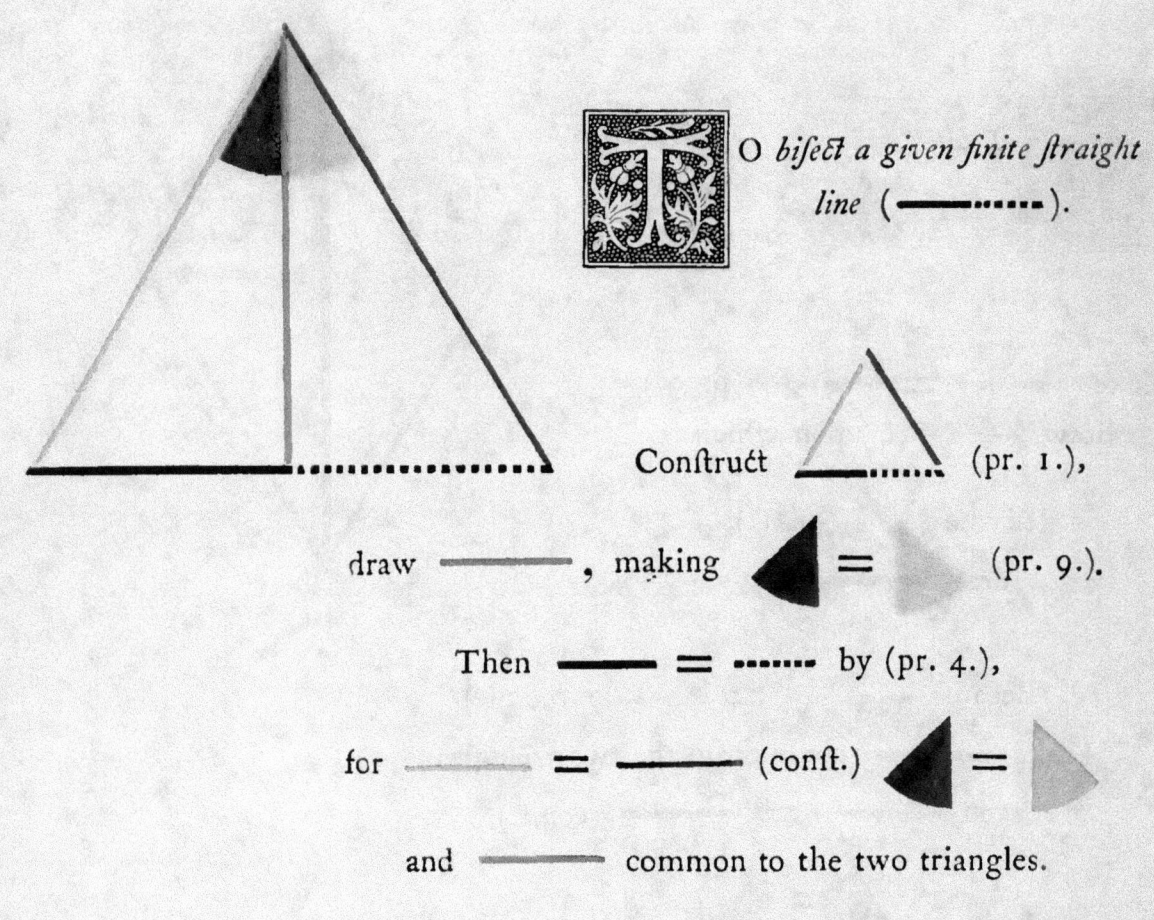

O *bifect a given finite ftraight line* (——— ·······).

Conftruct (pr. 1.),

draw ———, making ◢ = ◣ (pr. 9.).

Then ——— = ······· by (pr. 4.),

for ——— = ——— (conft.) ◢ = ◣

and ——— common to the two triangles.

Therefore the given line is bifected.

Q. E. D.

ROM *a given point* (———), *in a given straight line* (——————), *to draw a perpendicular.*

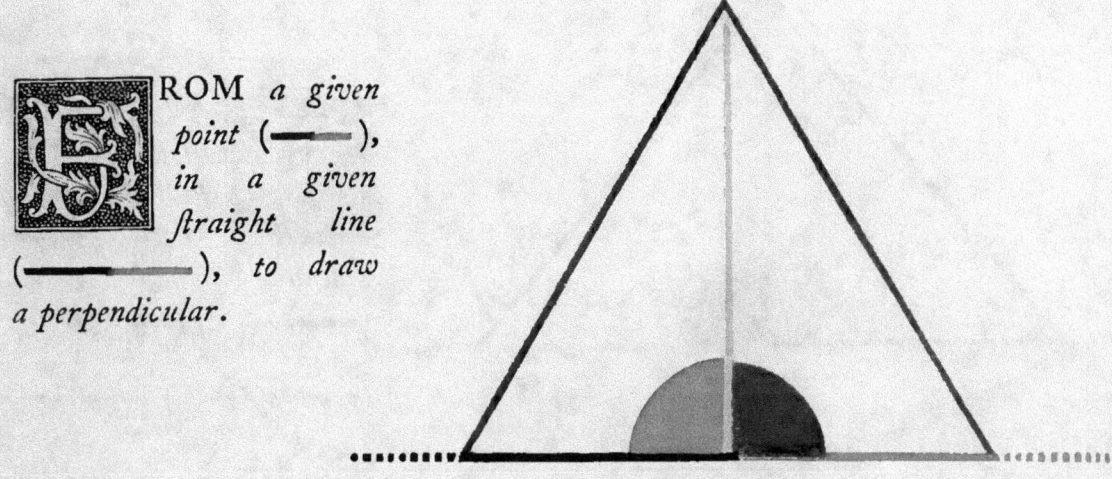

Take any point (——— ····) in the given line,

cut off ——— = ——— (pr. 3.),

construct △ (pr. 1.),

draw ——— and it fhall be perpendicular to the given line.

For ——— = ——— (conft.)

 = ——— (conft.)

and ——— common to the two triangles.

Therefore ◗ = ◗ (pr. 8.)

∴ ——— ⊥ ——— (def. 10.).

Q. E. D.

O draw a
ſtraight line
perpendicular
to a given
indefinite ſtraight line
(━━━━━) *from a given*

(*point* △) *without.*

With the given point △ as centre, at one ſide of the
line, and any diſtance ━━━━━ capable of extending to
the other ſide, deſcribe ‿ ,

Make ━━━━━ ═ ━━━━━ (pr. 10.)

draw ━━━━━ , ━━━━━ and ━━━━━ .

then ━━━━━ ⊥ ━━━━━ .

For (pr. 8.) ſince ━━━━━ ═ ━━━━━ (conſt.)

━━━━━ common to both,

and ━━━━━ ═ ━━━━━ (def. 15.)

∴ ◗ ═ ◗ , and

∴ ━━━━━ ⊥ ━━━━━ (def. 10.).

Q. E. D.

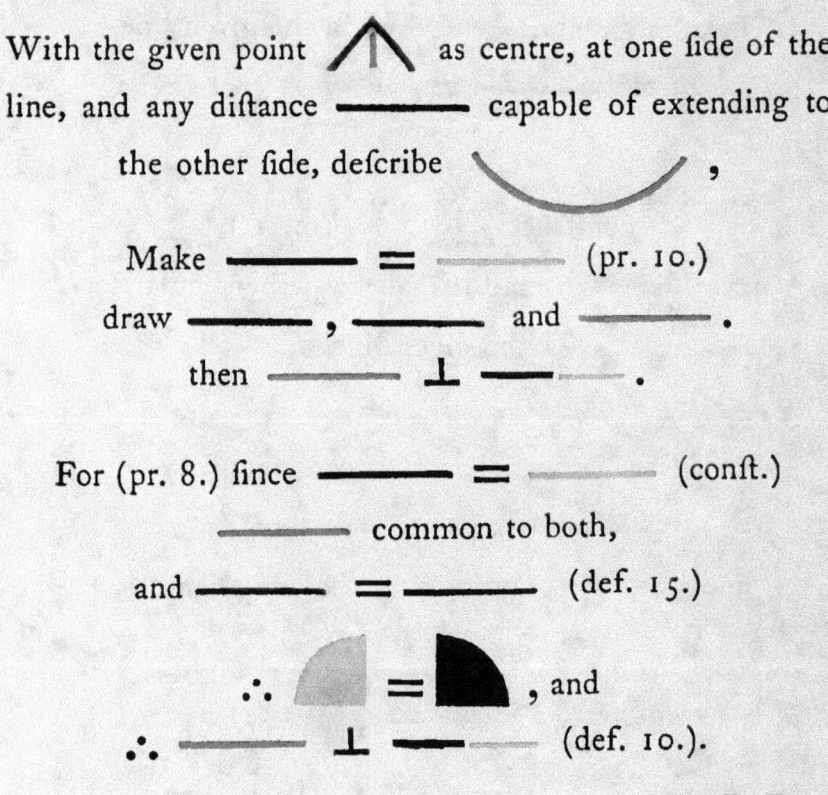

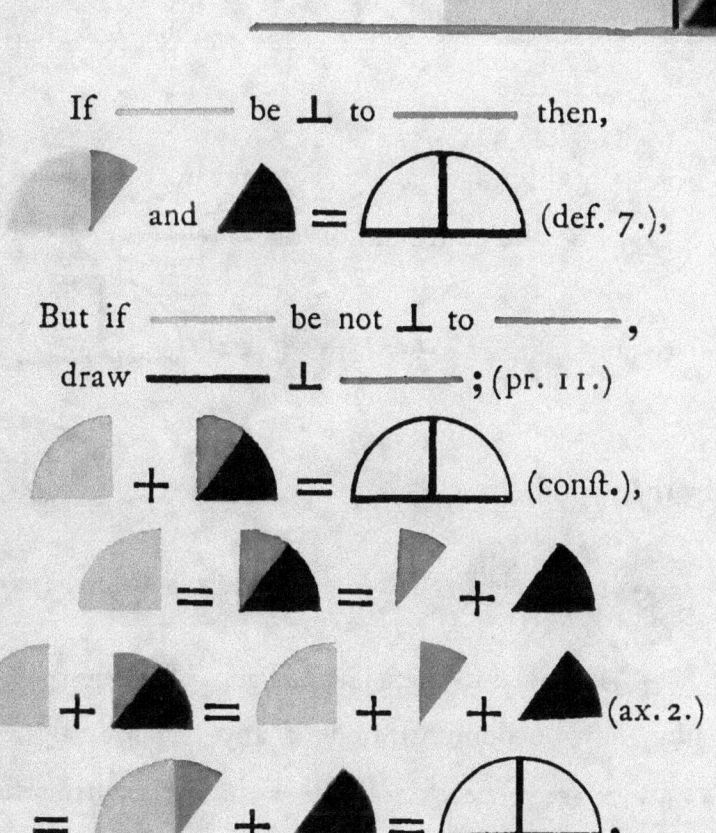

HEN *a straight line* (——) *standing upon another straight line* (————) *makes angles with it; they are either two right angles or together equal to two right angles.*

If ——— be ⊥ to ——— then,

and ▲ = (def. 7.),

But if ——— be not ⊥ to ———,

draw ——— ⊥ ———; (pr. 11.)

+ = (conft.),

= = +

∴ + = + + (ax. 2.)

= + = .

Q. E. D.

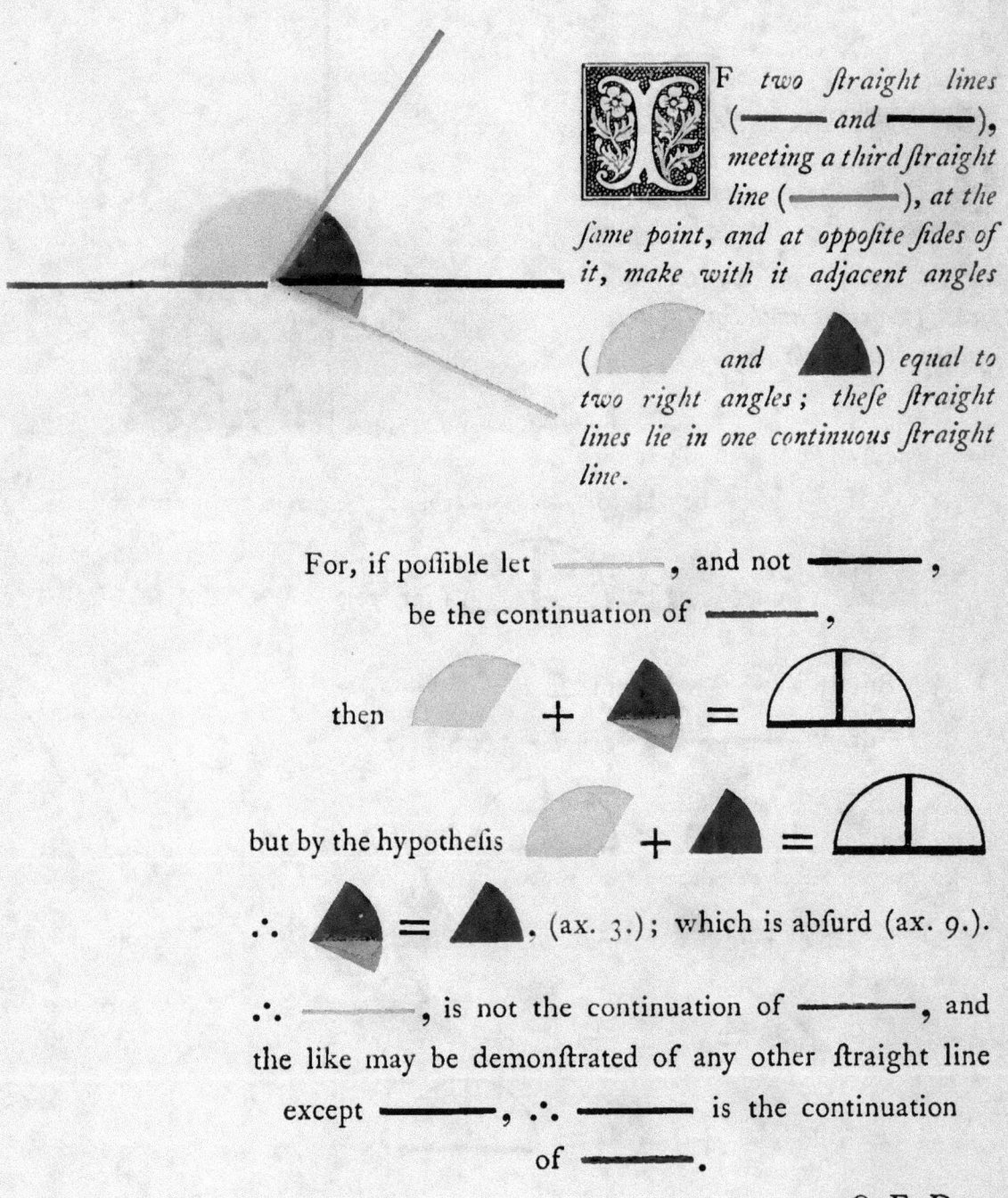

I F *two straight lines* (—— *and* ——), *meeting a third straight line* (——), *at the same point, and at opposite sides of it, make with it adjacent angles* (▨ *and* ◣) *equal to two right angles; these straight lines lie in one continuous straight line.*

For, if possible let ——, and not ——, be the continuation of ——,

then ◨ + ◣ = ◗

but by the hypothesis ◨ + ◣ = ◗

∴ ◣ = ◣ , (ax. 3.); which is absurd (ax. 9.).

∴ ——, is not the continuation of ——, and the like may be demonstrated of any other straight line except ——, ∴ —— is the continuation of ——.

Q. E. D.

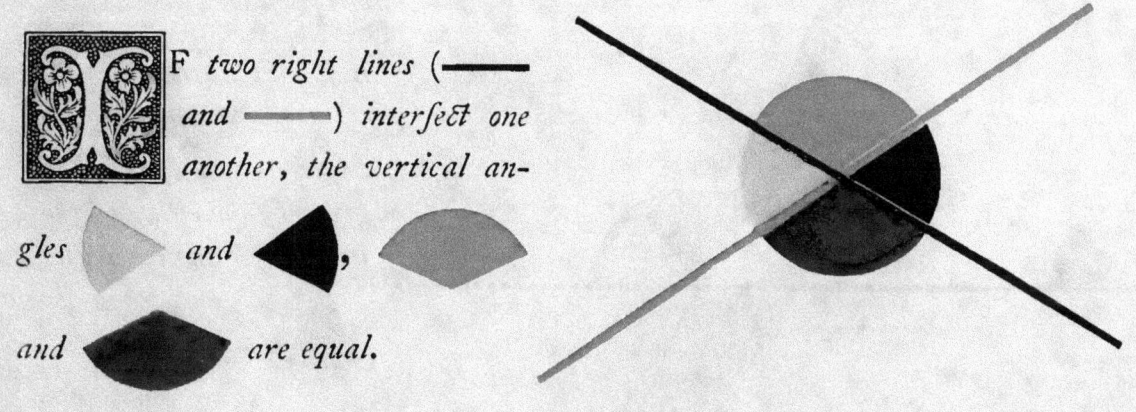

F *two right lines* (———— *and* ————) *interſect one another, the vertical angles* and , and are equal.

∴ =

In the ſame manner it may be ſhown that

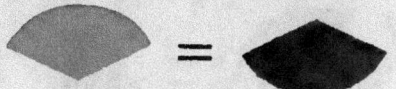

Q. E. D.

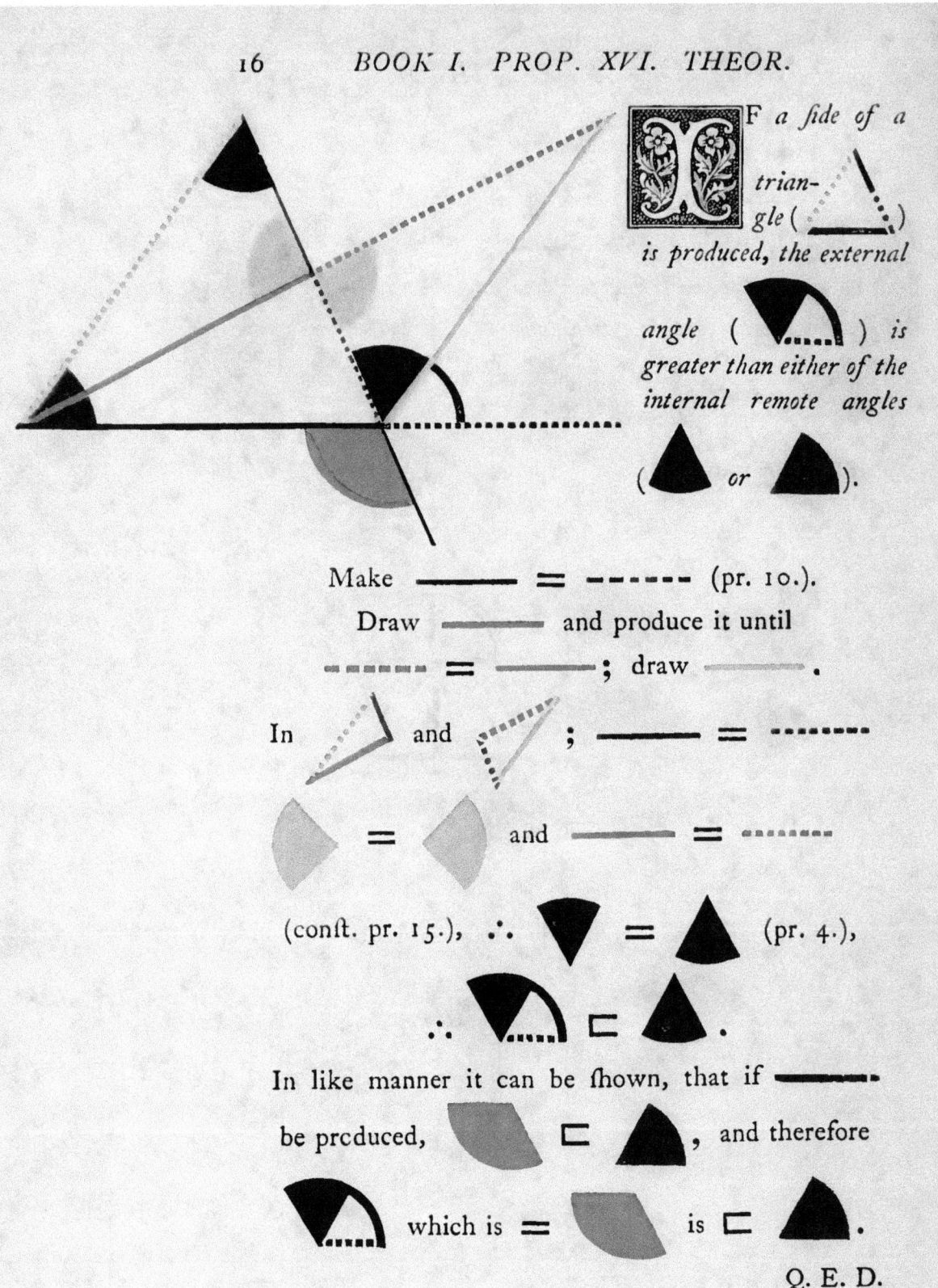

F a fide of a triangle ( ) is produced, the external

angle ( ) is greater than either of the internal remote angles

(or).

Make ——— = - - - - - - (pr. 10.).

Draw ——— and produce it until

- - - - - - = ——— ; draw ———— .

In and ; ——— = - - - - - - -

= and ——— = - - - - - - -

(conſt. pr. 15.), ∴ = (pr. 4.),

∴ ⊏ .

In like manner it can be ſhown, that if ———

be prcduced, ⊏ , and therefore

which is = is ⊏ .

Q. E. D.

NY *two angles of a tri-*  *angle are to-gether lefs than two right angles.*

Produce ————, then will

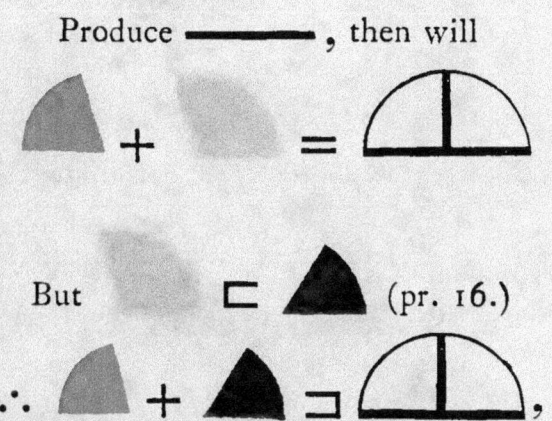

But ⊏ ▲ (pr. 16.)

∴ + ▲ ⊐ ,

and in the fame manner it may be fhown that any other two angles of the triangle taken together are lefs than two right angles.

Q. E. D.

D

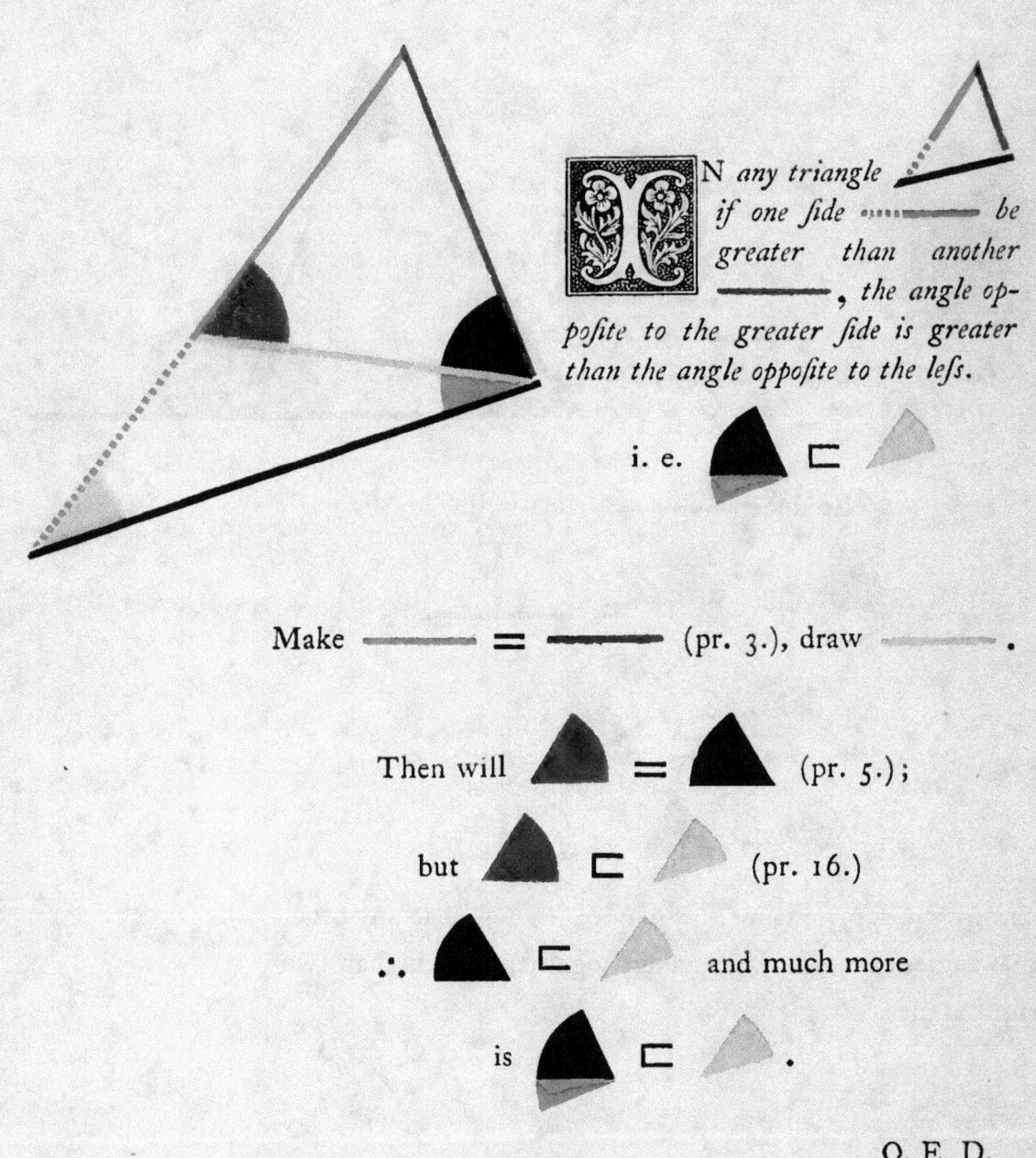

N any triangle if one fide ········ be greater than another ———, the angle oppofite to the greater fide is greater than the angle oppofite to the lefs.

i. e.

Make ——— = ——— (pr. 3.), draw ———.

Then will ▲ = ▲ (pr. 5.);

but ◣ ⊏ ◣ (pr. 16.)

∴ ◣ ⊏ ◣ and much more

is ◣ ⊏ ◣ .

Q. E. D.

18

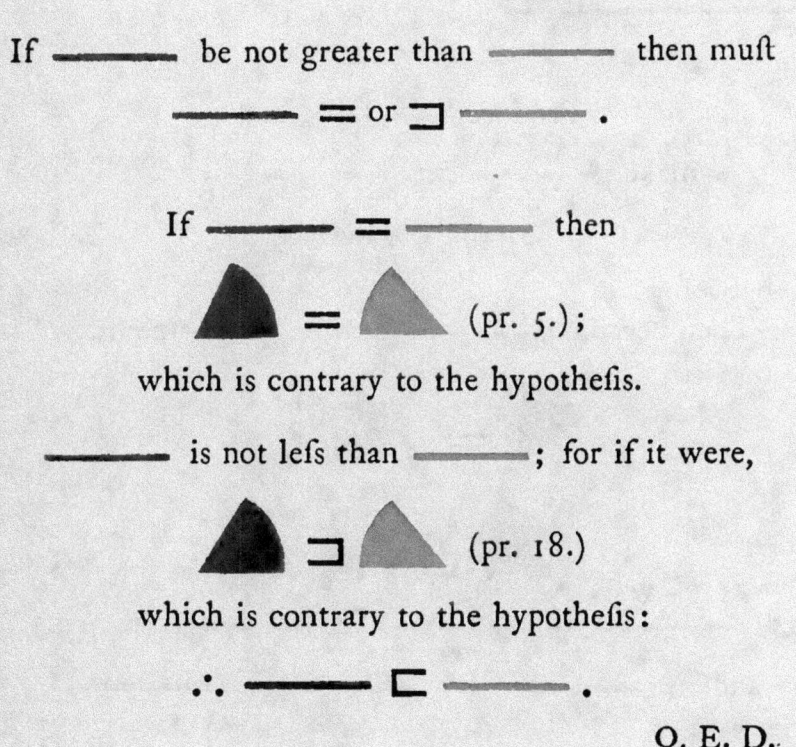

I F *in any triangle* one *angle* ▲ *be greater than another* ▲ *the ſide* ▬ *which is opposite to the greater angle, is greater than the ſide* ▬ *opposite the leſs.*

If ▬ be not greater than ▬ then muſt

▬ = or ⊐ ▬ .

If ▬ = ▬ then

▲ = ▲ (pr. 5.);

which is contrary to the hypotheſis.

▬ is not leſs than ▬ ; for if it were,

▲ ⊐ ▲ (pr. 18.)

which is contrary to the hypotheſis:

∴ ▬ ⊏ ▬ .

Q. E. D.

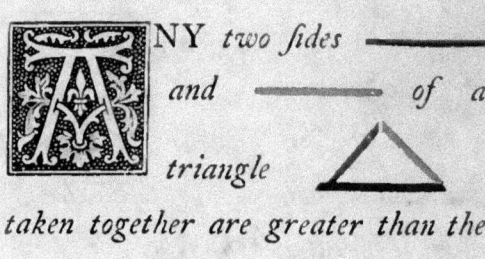

NY *two sides* —————— *and* ————— *of a triangle* *taken together are greater than the third side* (——).

Produce ———, and
make ------ = ———— (pr. 3.);
draw ————.

Then becauſe ------ = ———— (conſt.),

◣ = ◤ (pr. 5.)

∴ ◣ ⊏ ◤ (ax. 9.)

∴ ——— + ------ ⊏ ——— (pr. 19.)

and ∴ ——— + ———— ⊏ ——— .

Q. E. D

F *from any point* (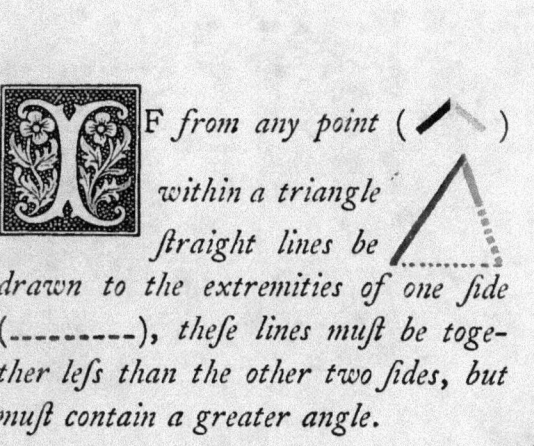) *within a triangle straight lines be drawn to the extremities of one side* (----------), *these lines must be together less than the other two sides, but must contain a greater angle.*

Produce ————,

———— + ———— ⊏ ———— ···· (pr. 20.),

add ----- to each,

———— + ——— ⊏ ——— + ----- (ax. 4.)

In the same manner it may be shown that

——···· + ----- ⊏ ———— + ———— , ∴

———— + ——— ⊏ ———— + ———— ,

which was to be proved.

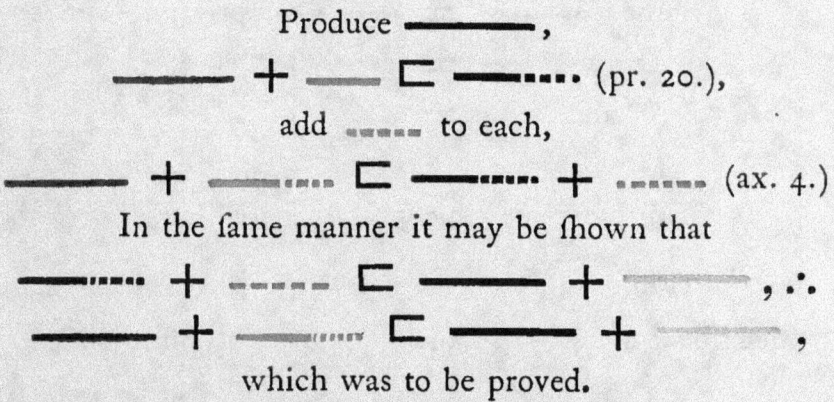

Again 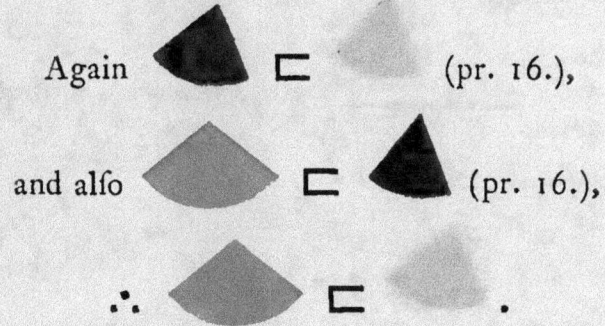 ⊏ (pr. 16.),

and also ⊏ (pr. 16.),

∴ ⊏ .

Q. E. D.

I. PROP. XXII. THEOR.

BOOK

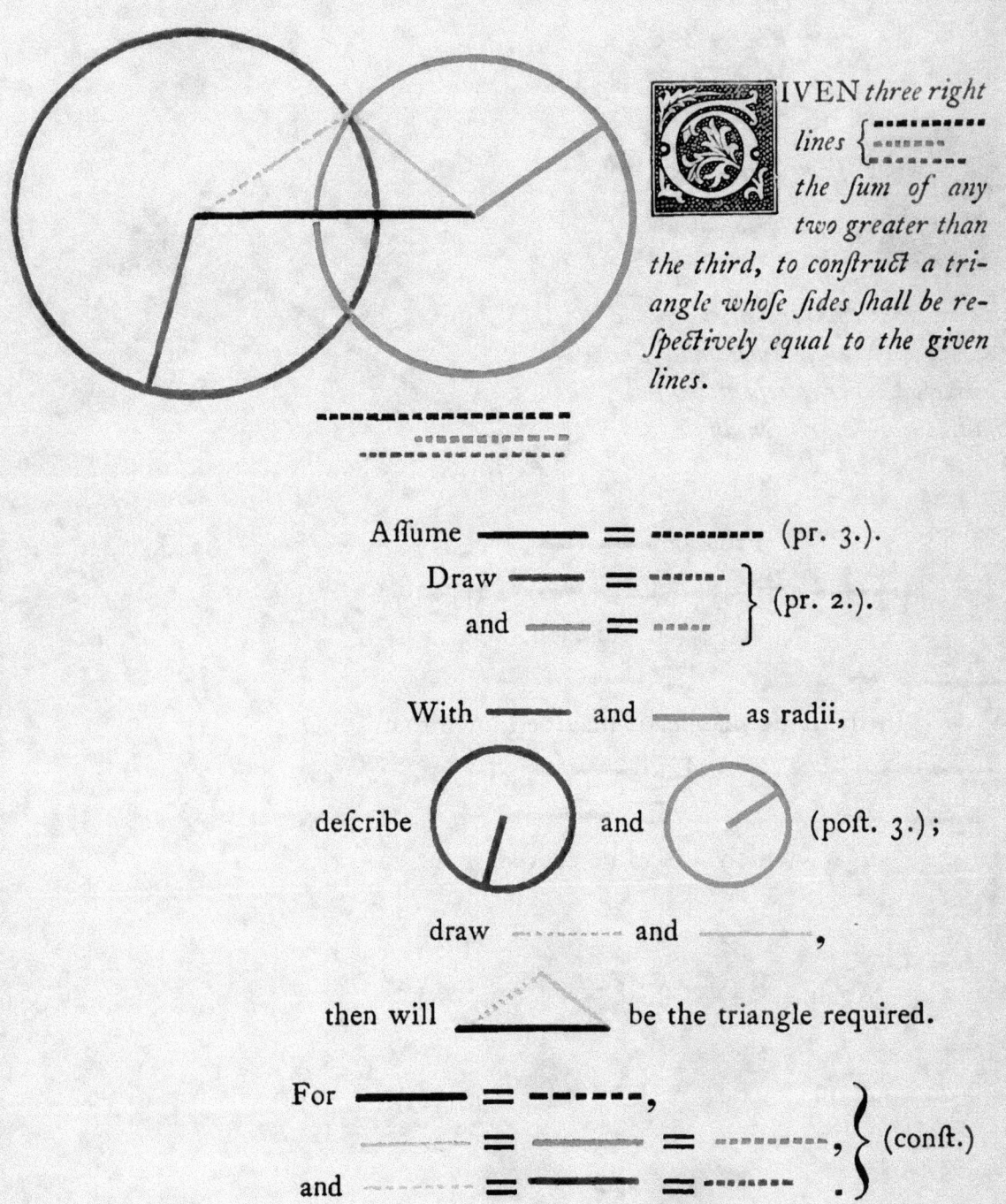

GIVEN *three right lines* { the *sum of any two greater than the third, to construct a triangle whose sides shall be respectively equal to the given lines.*

Assume ——— = ·········· (pr. 3.).

Draw ——— = ······· } (pr. 2.).
and ——— = ·····

With ——— and ——— as radii,

describe and (post. 3.);

draw ········· and ———,

then will be the triangle required.

For ——— = ·······,

——— = ——— = ········· , } (const.)

and ——— = ——— = ······· .

Q. E. D.

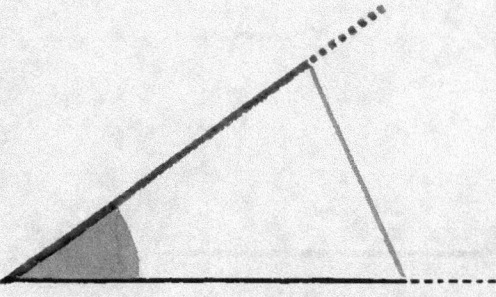

T *a given point* (——) *in a given straight line* (——·····), *to make an angle equal to a given rectilineal angle* (◢).

Draw ———— between any two points in the legs of the given angle.

Construct ◢ (pr. 22.) so that ▬▬▬ = ————, ———— = ——— and ▬▬▬ ——— .

Then ◢ = ◢ (pr. 8.).

Q. E. D.

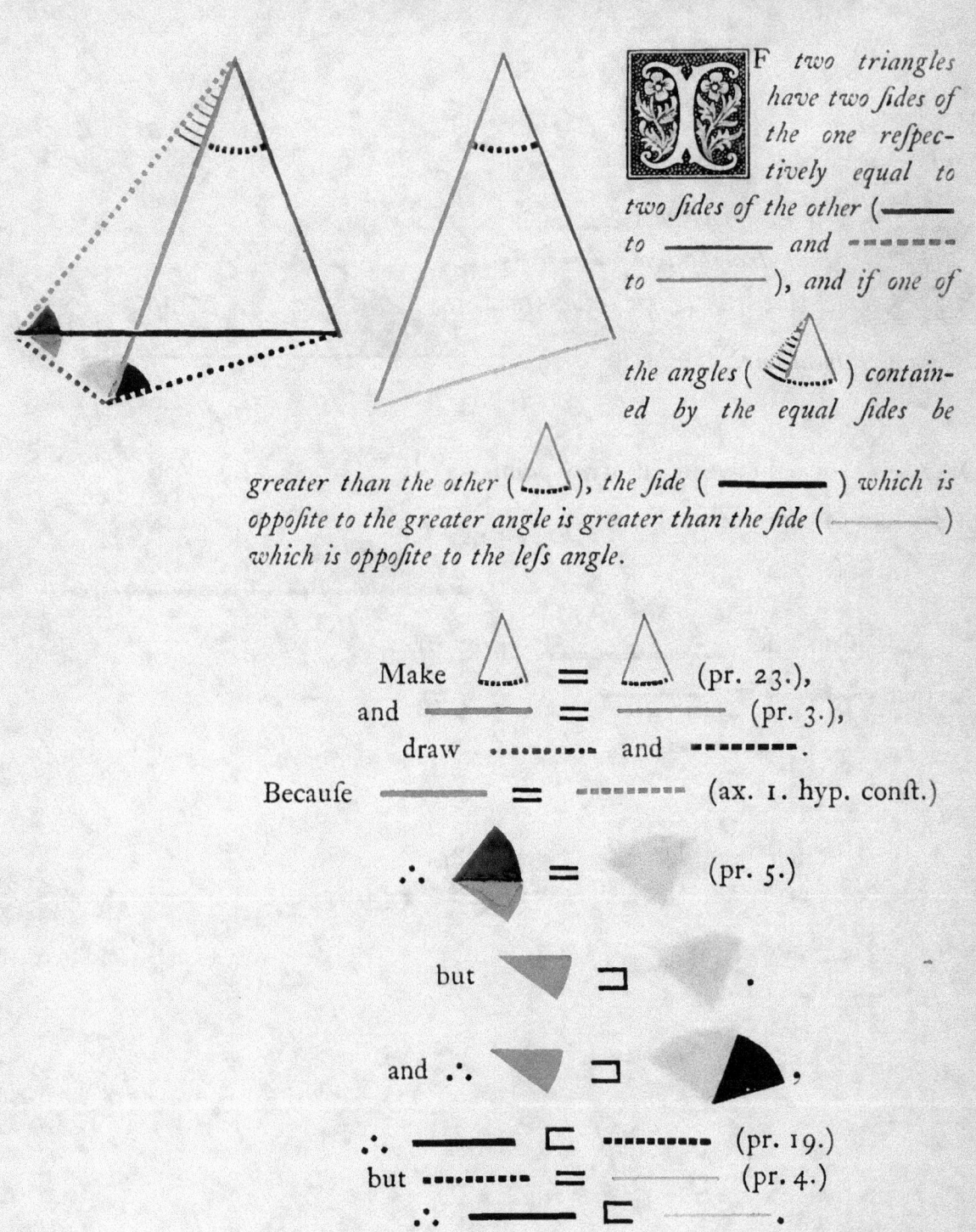

I F two triangles have two sides of the one respectively equal to two sides of the other (——— to ——— and ------- to ———), and if one of the angles ( ) contained by the equal sides be greater than the other ( ), the side (———) which is opposite to the greater angle is greater than the side (———) which is opposite to the less angle.

Make ⚬ = ⚬ (pr. 23.),
and ——— = ——— (pr. 3.),
draw .... and ----------.
Because ——— = ------- (ax. 1. hyp. conft.)

∴ = (pr. 5.)

but ⊐ .

and ∴ ⊐ ,

∴ ——— ⊏ ------- (pr. 19.)
but .... = ——— (pr. 4.)
∴ ——— ⊏ .

Q. E. D.

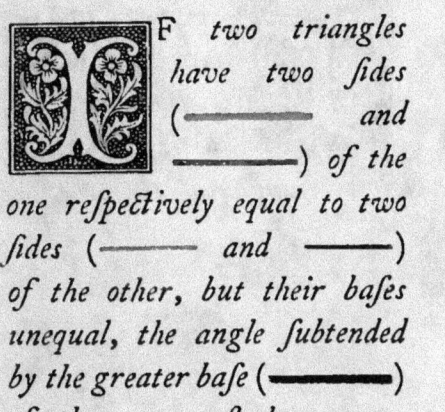

F two triangles have two sides (━━━ and ━━━) of the one respectively equal to two sides (━━━ and ━━━) of the other, but their bases unequal, the angle subtended by the greater base (━━━━) of the one, must be greater than the angle subtended by the less base (━━━━) of the other.

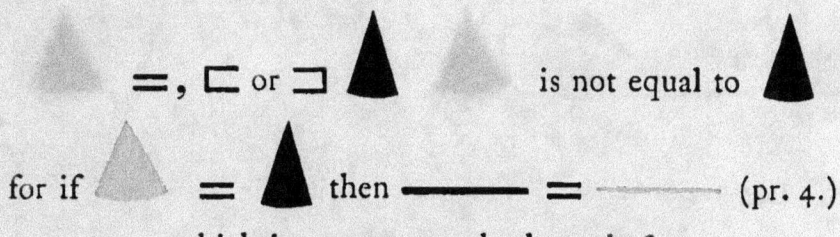

▲ =, ⊏ or ⊐ ▲   ▲ is not equal to ▲

for if ▲ = ▲ then ━━━ = ━━━ (pr. 4.)

which is contrary to the hypothefis;

▲ is not lefs than ▲

for if ▲ ⊐ ▲

then ━━━ ⊐ ━━━ (pr. 24.),

which is alfo contrary to the hypothefis:

∴ 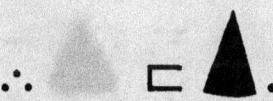 ▲ ⊏ ▲ .

Q. E. D.

E

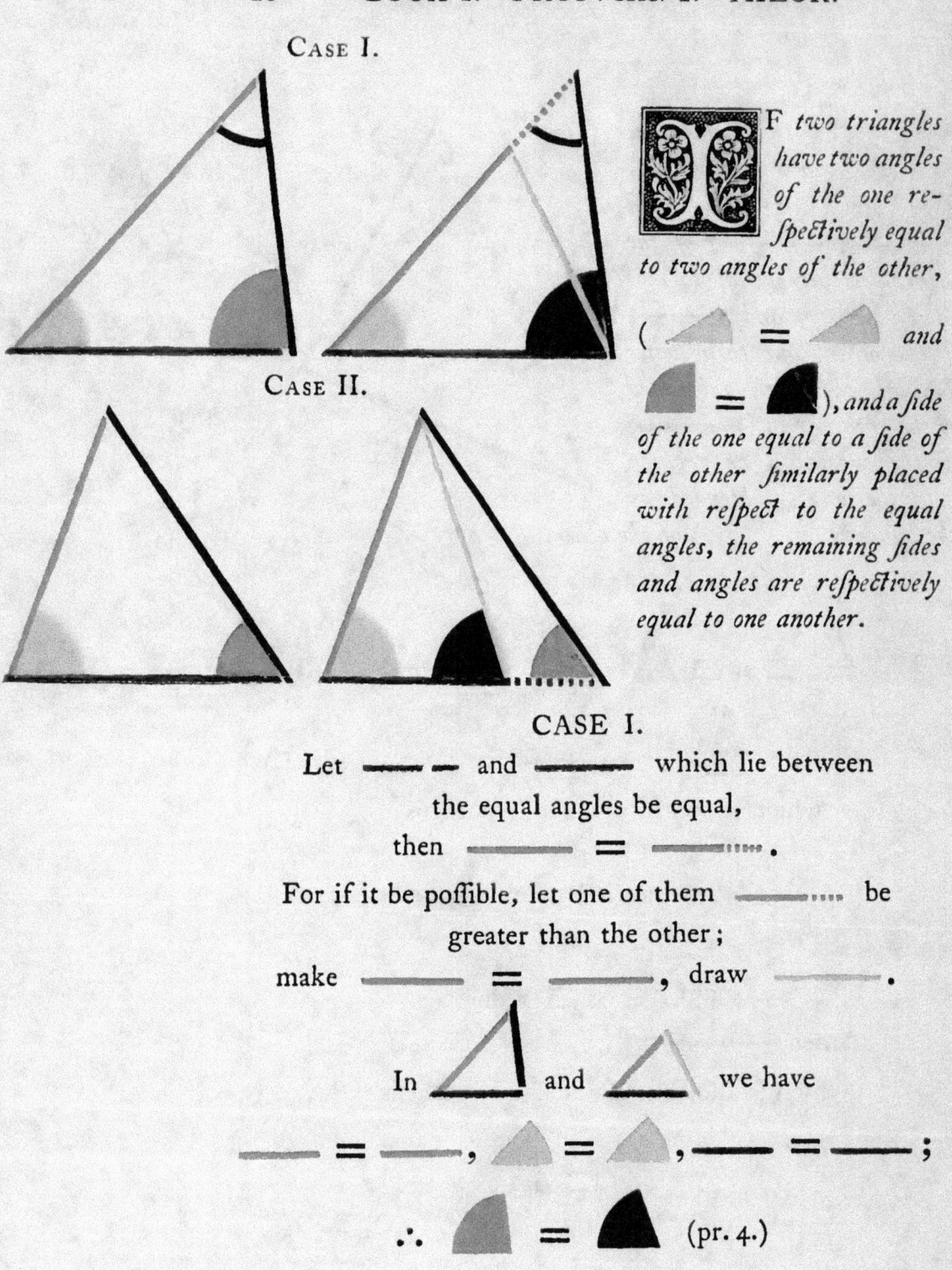

CASE I.

CASE II.

F *two triangles have two angles of the one respectively equal to two angles of the other,*

(▰ = ▰ *and* ◢ = ◢), *and a side of the one equal to a side of the other similarly placed with respect to the equal angles, the remaining sides and angles are respectively equal to one another.*

CASE I.

Let ━━ and ━━ which lie between
the equal angles be equal,

then ━━ = ━━ .

For if it be possible, let one of them ━━ be
greater than the other;

make ━━ = ━━ , draw ━━ .

In ◸ and ◺ we have

━━ = ━━ , ◤ = ◤ , ━━ = ━━ ;

∴ ◗ = ◗ (pr. 4.)

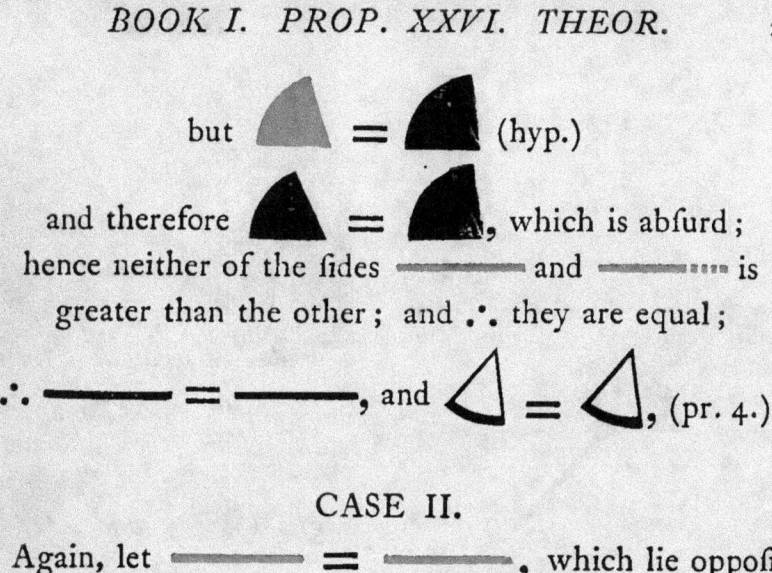

but ▰ = ◣ (hyp.)

and therefore ◣ = ◣, which is abfurd;
hence neither of the fides ▬▬ and ▬▬···· is
greater than the other; and ∴ they are equal;

∴ ▬▬ = ▬▬, and ◁ = ◁, (pr. 4.).

CASE II.

Again, let ▬▬ = ▬▬, which lie oppofite

the equal angles ▲ and ▲. If it be poffible, let
▬▬···· ⊏ ▬▬, then take ▬▬ = ▬▬,
draw ▬▬.

Then in ◿ and ◿ we have ▬▬ = ▬▬,

▬▬ = ▬▬ and ◣ = ◣,

∴ ◣ = ◣ (pr. 4.)

but ◣ = ◣ (hyp.)

∴ ◣ = ◣ which is abfurd (pr. 16.).

Confequently, neither of the fides ▬▬ or ▬▬···· is
greater than the other, hence they muft be equal. It
follows (by pr. 4.) that the triangles are equal in all
refpects.

Q. E. D.

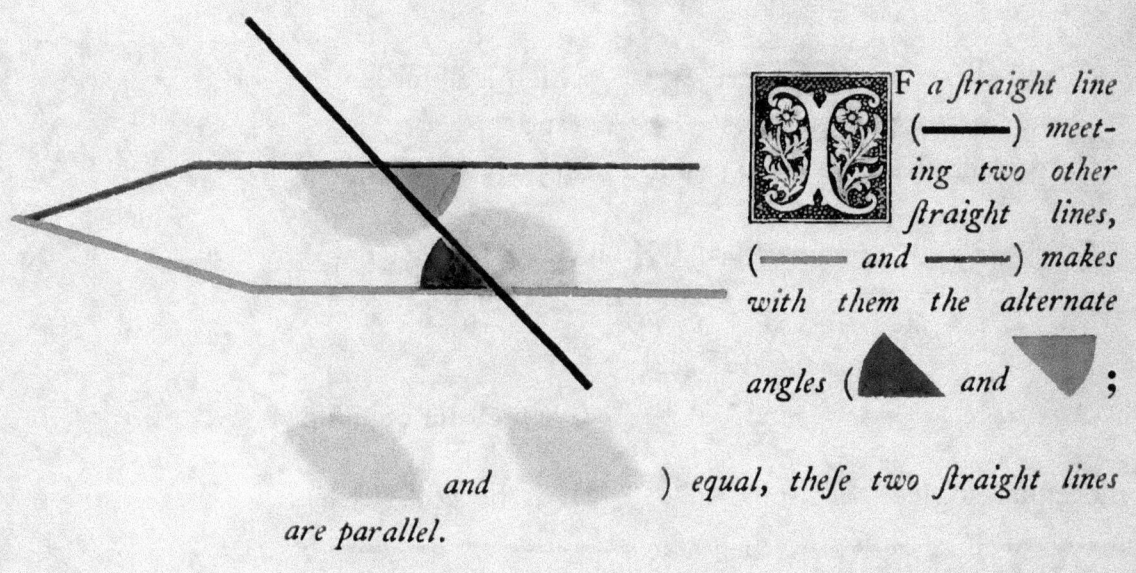

I F *a straight line* (——) *meeting two other straight lines,* (—— and ——) *makes with them the alternate angles* (◣ *and* ◥ ; *and*) *equal, thefe two ftraight lines are parallel.*

If —— be not parallel to —— they fhall meet when produced.

If it be poffible, let thofe lines be not parallel, but meet when produced; then the external angle ◥ is greater than ◣ (pr. 16), but they are alfo equal (hyp.), which is abfurd: in the fame manner it may be fhown that they cannot meet on the other fide; ∴ they are parallel.

Q. E. D.

F *a straight line* (——), *cutting two other straight lines* (—— *and* ——), *makes the external equal to the internal and opposite angle, at the same side of the cutting line (namely,*

), or if it makes the two internal angles at the same side (*and* *, or* *and* *) together equal to two right angles, those two straight lines are parallel.*

First, if $=$, then $=$ (pr. 15.),

$\therefore$ $=$ $\therefore$ —— $\parallel$ —— (pr. 27.).

Secondly, if $+$ $=$,

then $+$ $=$ (pr. 13.),

$\therefore$ $+$ $=$ $+$ (ax. 3.)

$\therefore$ $=$

$\therefore$ —— $\parallel$ —— (pr. 27.)

Q. E. D.

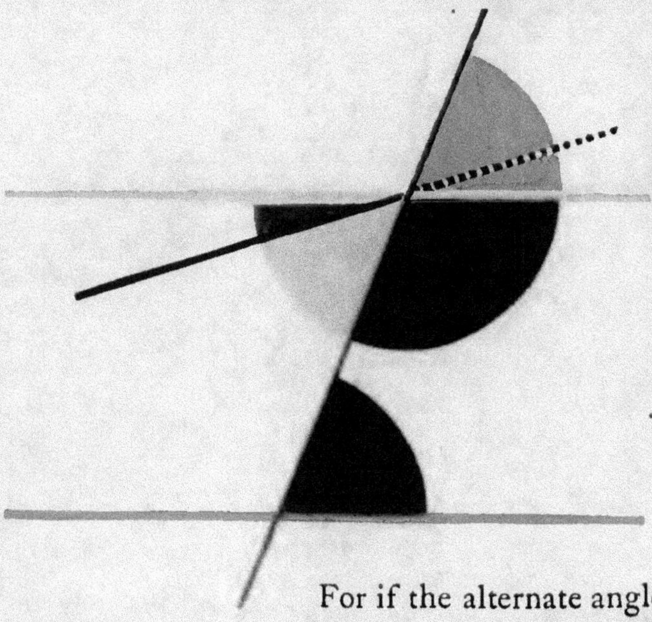

STRAIGHT *line* (———) *falling on two parallel ſtraight lines (* 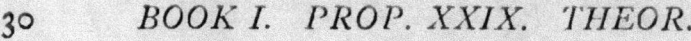 *and* ———)*, makes the alternate angles equal to one another ; and alſo the external equal to the internal and oppoſite angle on the ſame ſide ; and the two internal angles on the ſame ſide together equal to two right angles.*

For if the alternate angles ▰ and ◣ be not equal,

draw ———, making ▽ = ◣ (pr. 23).
Therefore ━━━━•••• ‖ ——— (pr. 27.) and therefore two ſtraight lines which interſect are parallel to the
ſame ſtraight line, which is impoſſible (ax. 12).

Hence the alternate angles ◣ and ◣ are not

unequal, that is, they are equal: ◣ = ◣ (pr. 15);

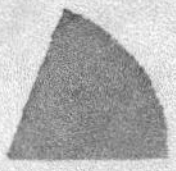

∴ ◣ = ◣, the external angle equal to the internal and oppoſite on the ſame ſide : if ◖ be added to

both, then ◣ + ◖ = ◖ = ⌓ (pr. 13).

That is to ſay, the two internal angles at the ſame ſide of
the cutting line are equal to two right angles.

Q. E. D.

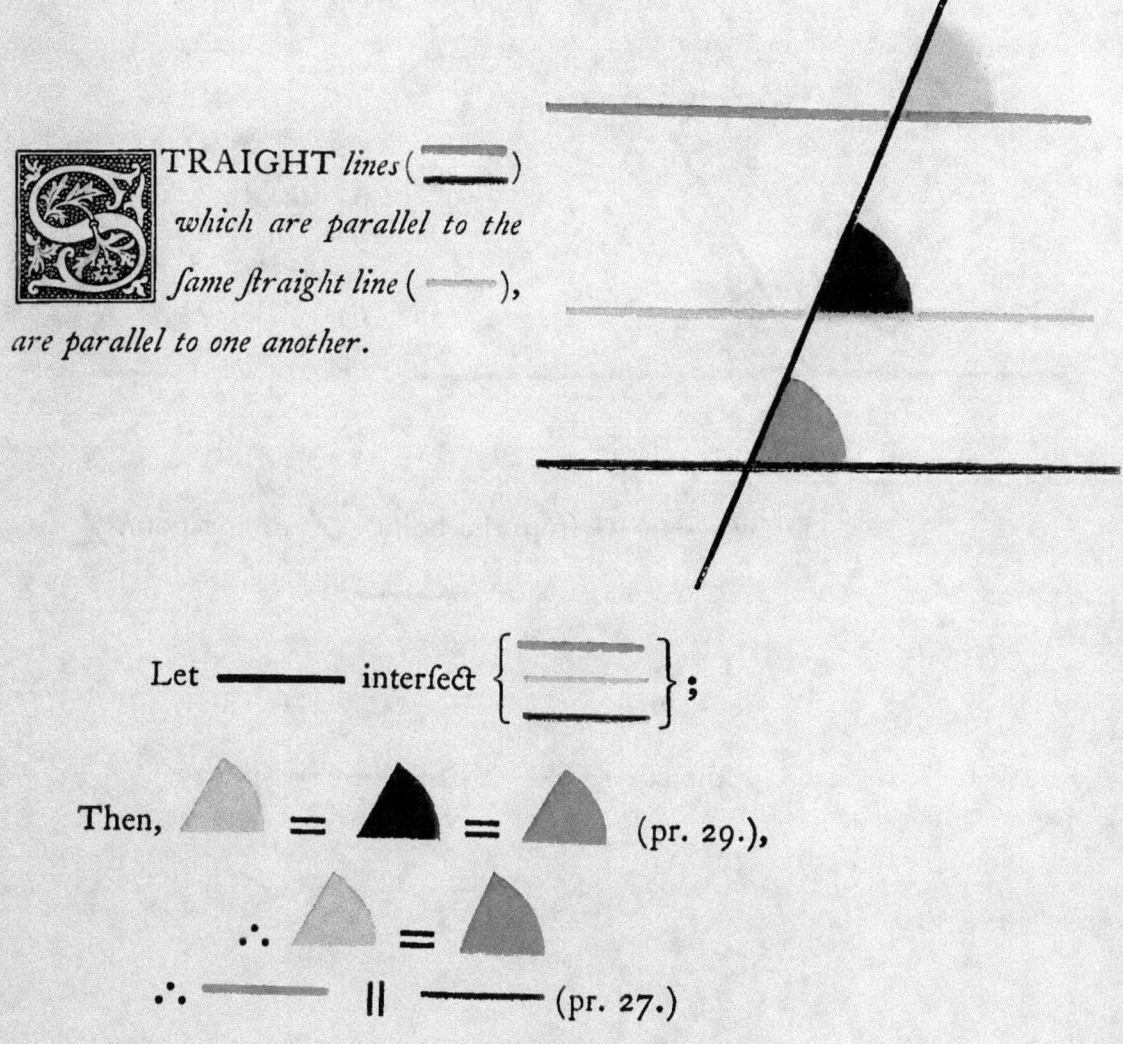

S TRAIGHT *lines* (—————)
which are parallel to the
same straight line (———),
are parallel to one another.

Let ——— interſect ⎰ ————— ⎱ ;

Then, ▲ = ▲ = ▲ (pr. 29.),

∴ ▲ = ▲

∴ ——— ‖ ——— (pr. 27.)

Q. E. D.

ROM *a given point* 7 *to draw a ſtraight line parallel to a given ſtraight line* (————).

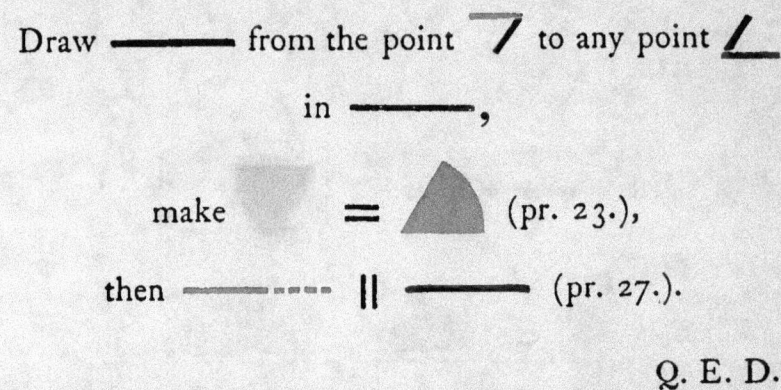

Draw ———— from the point 7 to any point ∠ in ————,

make ▽ = ◣ (pr. 23.),

then ———- - - ‖ ———— (pr. 27.).

Q. E. D.

F any *fide* (——————) of a triangle be pro-duced, the external

angle (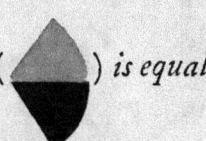) is equal

to the *fum* of the two internal and

oppofite angles (and), and the three internal angles of every triangle taken together are equal to two right angles.

Through the point draw

—— || —— (pr. 31.).

Then $\Big\{$ = $\Big\}$ (pr. 29.),

=

∴ + = (ax. 2.),

and therefore

 + + = = (pr. 13.).

Q. E. D.

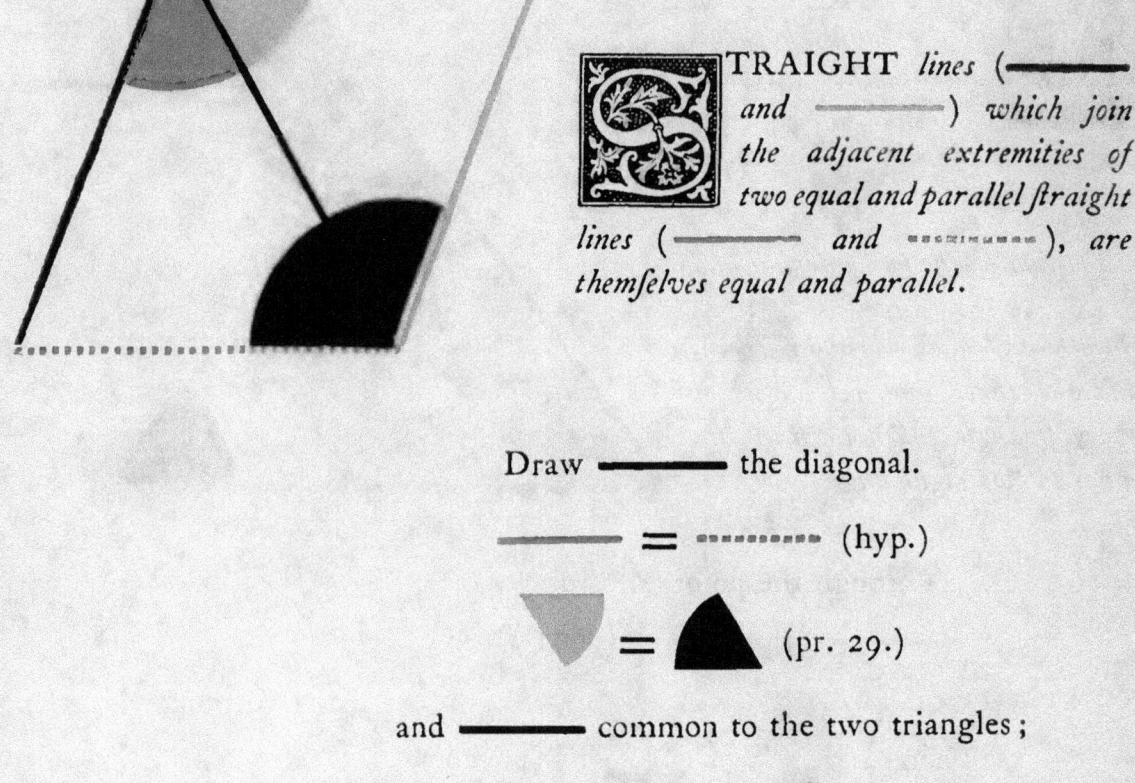

TRAIGHT *lines* (————— and —————) *which join the adjacent extremities of two equal and parallel straight lines* (————— *and* ━━━━━), *are themselves equal and parallel.*

Draw ————— the diagonal.

————— = ━━━━━ (hyp.)

▼ = ▲ (pr. 29.)

and ————— common to the two triangles;

∴ ————— = —————, and ▼ = ▲ (pr. 4.);

and ∴ ————— ‖ ————— (pr. 27.).

Q. E. D.

 HE *oppofite fides and angles of any parallelogram are equal, and the diagonal (————) divides it into two equal parts.*

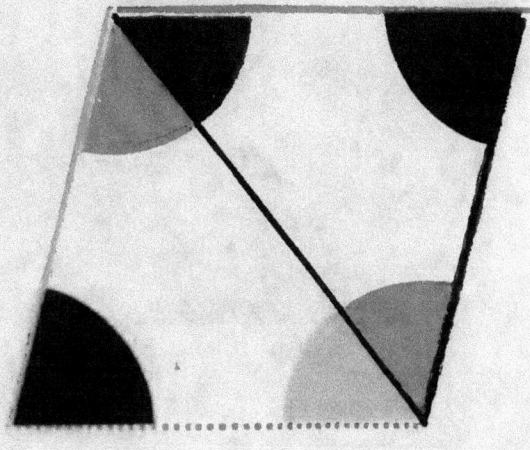

Since 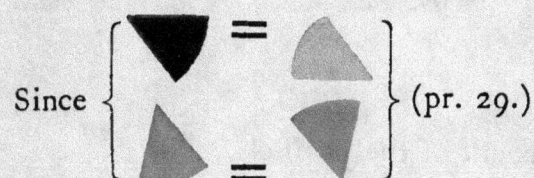 (pr. 29.)

and ———— common to the two triangles.

∴ (pr. 26.)

and = (ax.):

Therefore the oppofite fides and angles of the parallelo-gram are equal: and as the triangles and are equal in every refpect (pr. 4,), the diagonal divides the parallelogram into two equal parts.

Q. E. D.

ARALLELOGRAMS *on the fame bafe, and between the fame parallels, are (in area) equal.*

On account of the parallels,

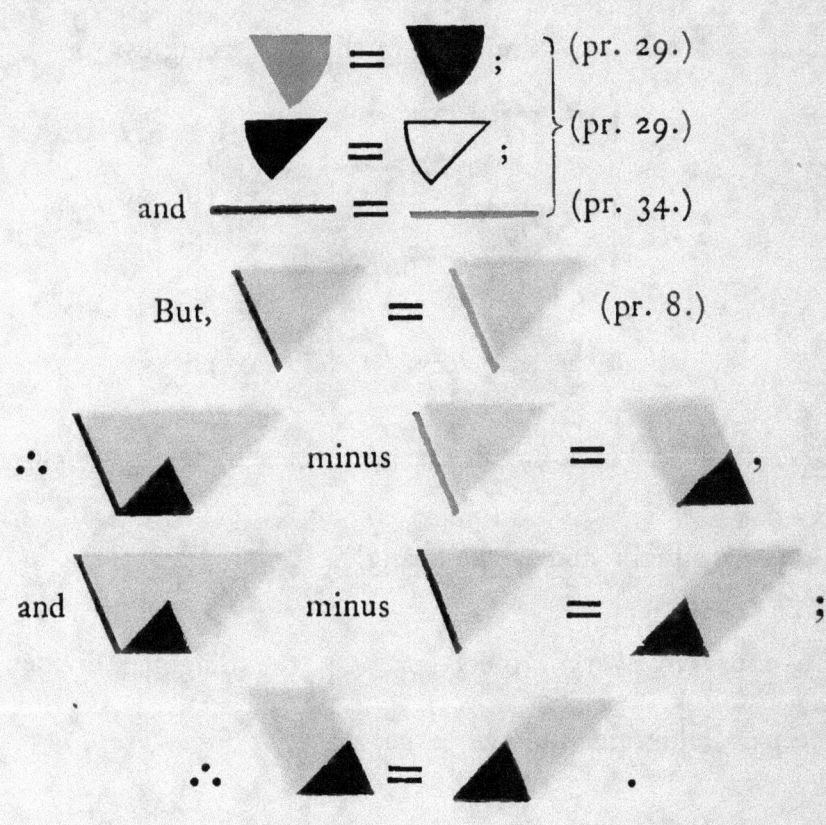

But,

∴

and

∴

Q. E. D.

 ARALLELO-
GRAMS

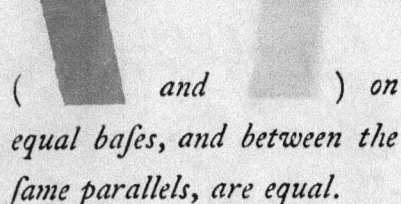

(*and*) *on equal bafes, and between the fame parallels, are equal.*

Draw ———— and — — — ,

——— = ——— = ——— , by (pr. 34, and hyp.);

∴ ——— = and ∥ ——— ;

∴ ——— = and ∥ — — — — (pr. 33.)

And therefore  is a parallelogram:

but 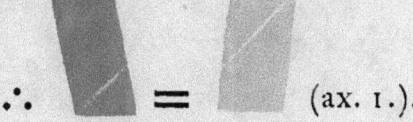 = = (pr. 35.)

∴ = (ax. 1.).

Q. E. D.

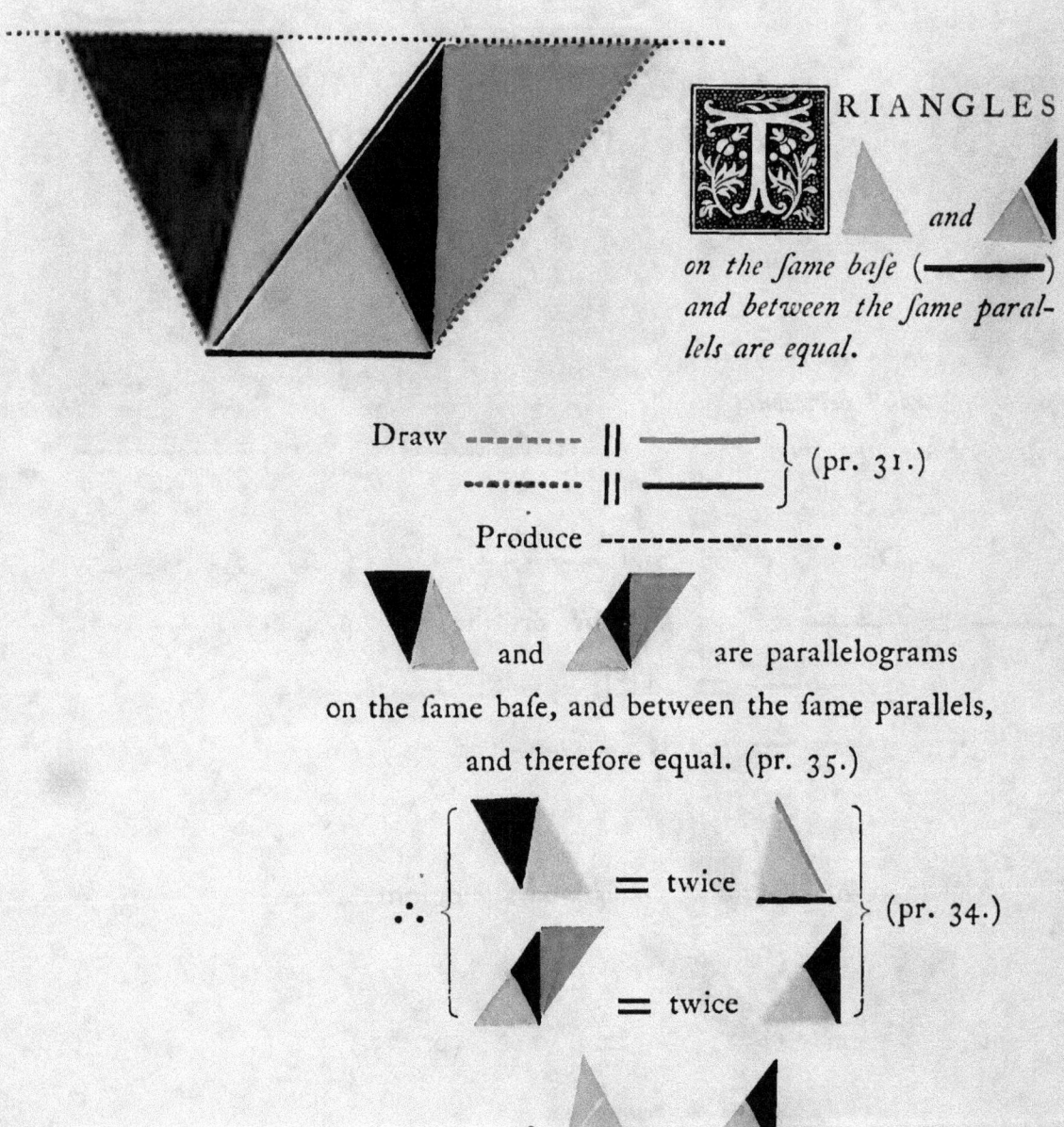

TRIANGLES and on the fame bafe (———) and between the fame parallels are equal.

Draw ⫽ } (pr. 31.)

Produce •

and are parallelograms on the fame bafe, and between the fame parallels, and therefore equal. (pr. 35.)

∴ { = twice = twice } (pr. 34.)

∴ =

Q. E D.

TRIANGLES

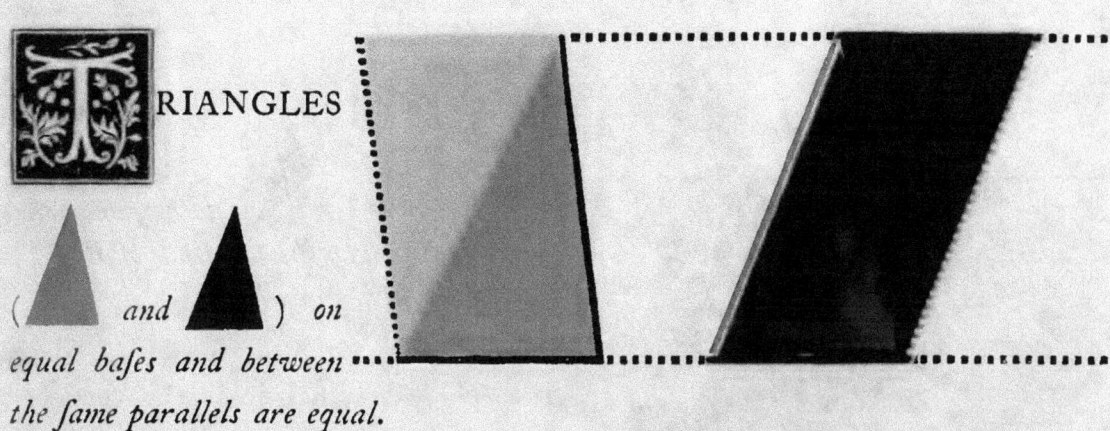

(and) on equal bafes and between the fame parallels are equal.

Draw ┅┅┅ ‖ ━━━ and ┄┄┄ ‖ ━━━ } (pr. 31.).

= ▰ (pr. 36.);

but = twice ▲ (pr. 34.),

and ▰ = twice ▲ (pr. 34.),

∴ ▲ = ▲ (ax. 7.).

Q. E. D.

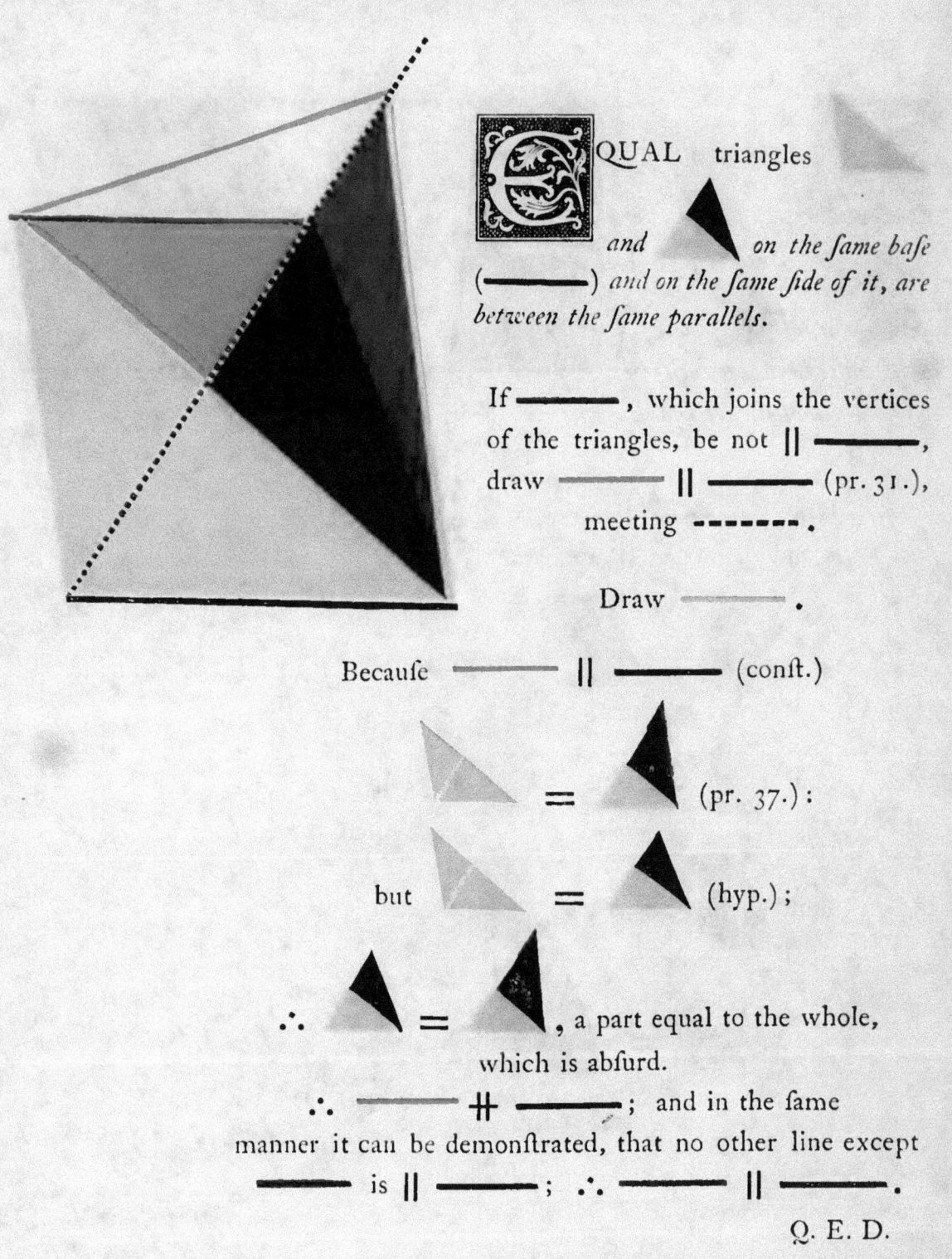

QUAL triangles and ▰ on the fame bafe (————) and on the fame fide of it, are between the fame parallels.

If ————, which joins the vertices of the triangles, be not ‖ ————, draw ———— ‖ ———— (pr. 31.), meeting - - - - - - .

Draw ————— .

Becaufe ————— ‖ ———— (conft.)

▰ = ▰ (pr. 37.):

but ▰ = ▰ (hyp.);

∴ ▰ = ▰ , a part equal to the whole, which is abfurd.

∴ ————— ╫ ————; and in the fame manner it can be demonftrated, that no other line except ———— is ‖ ————; ∴ ———— ‖ ————.

Q. E. D.

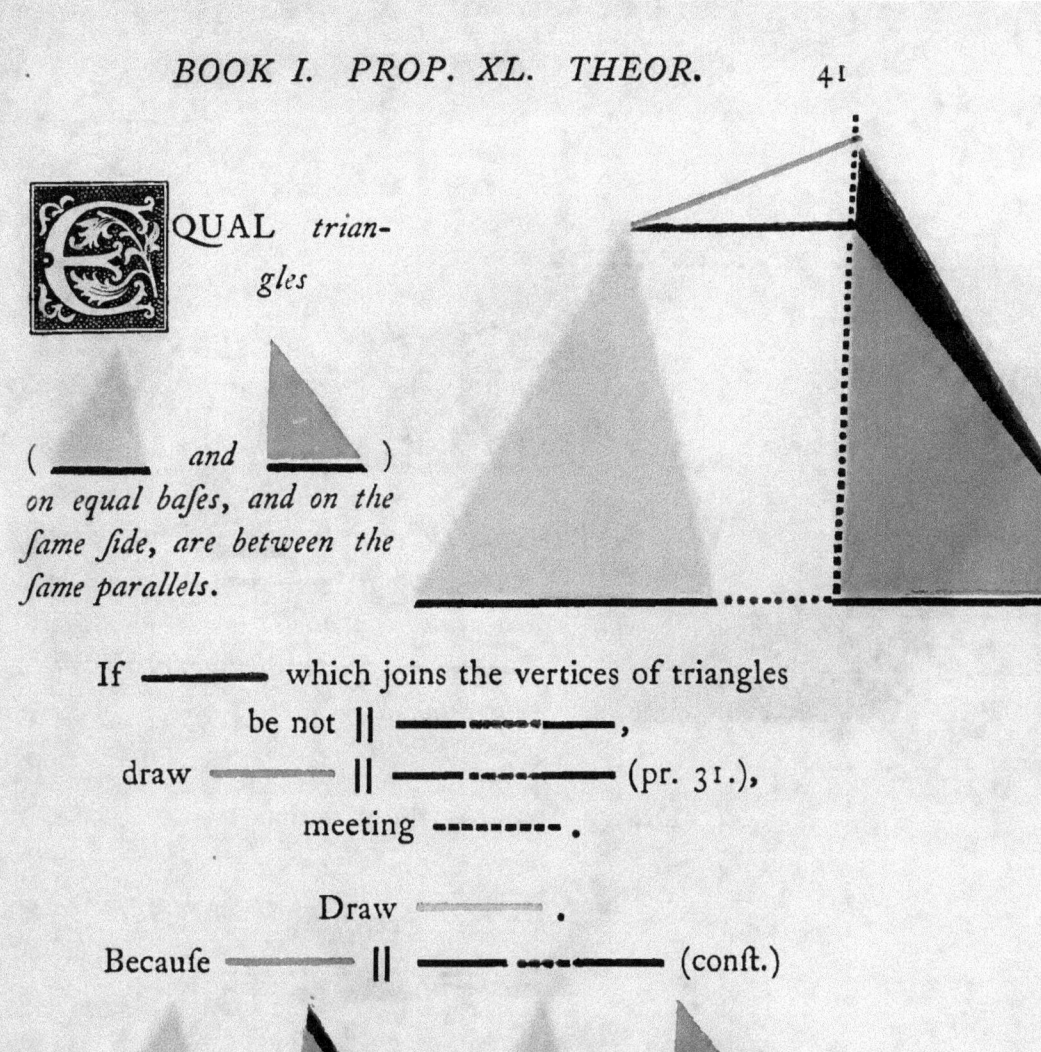

QUAL *trian-gles*

(_____ and _____) on equal baſes, and on the ſame ſide, are between the ſame parallels.

If _____ which joins the vertices of triangles be not ‖ _____,

draw _____ ‖ _____ (pr. 31.),

meeting _____.

Draw _____.

Becauſe _____ ‖ _____ (conſt.)

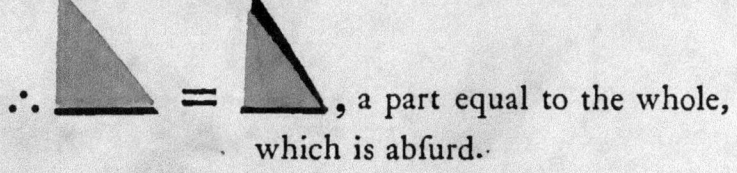

_____ = ◺ but _____ = ◺

∴ ◺ = ◺, a part equal to the whole, which is abſurd.

∴ _____ ‖‖ _____ : and in the ſame manner it can be demonſtrated, that no other line except

_____ is ‖ _____ : ∴ _____ ‖ _____.

Q. E. D.

G

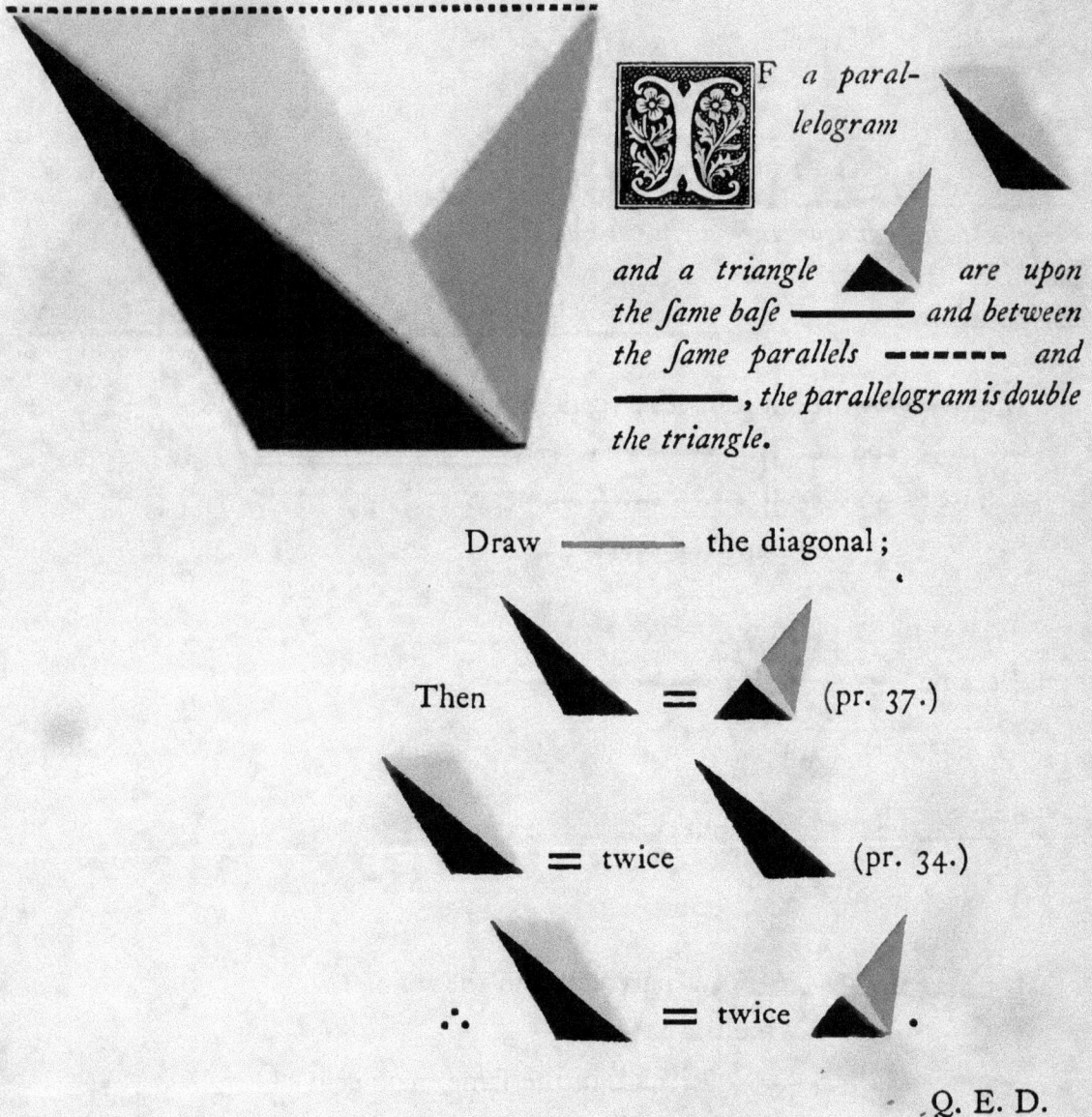

F a paral-
lelogram

and a triangle   are upon
the same base ——— and between
the same parallels ------ and
———, the parallelogram is double
the triangle.

Draw ——— the diagonal;

Then   =   (pr. 37.)

= twice   (pr. 34.)

∴   = twice   .

Q. E. D.

42

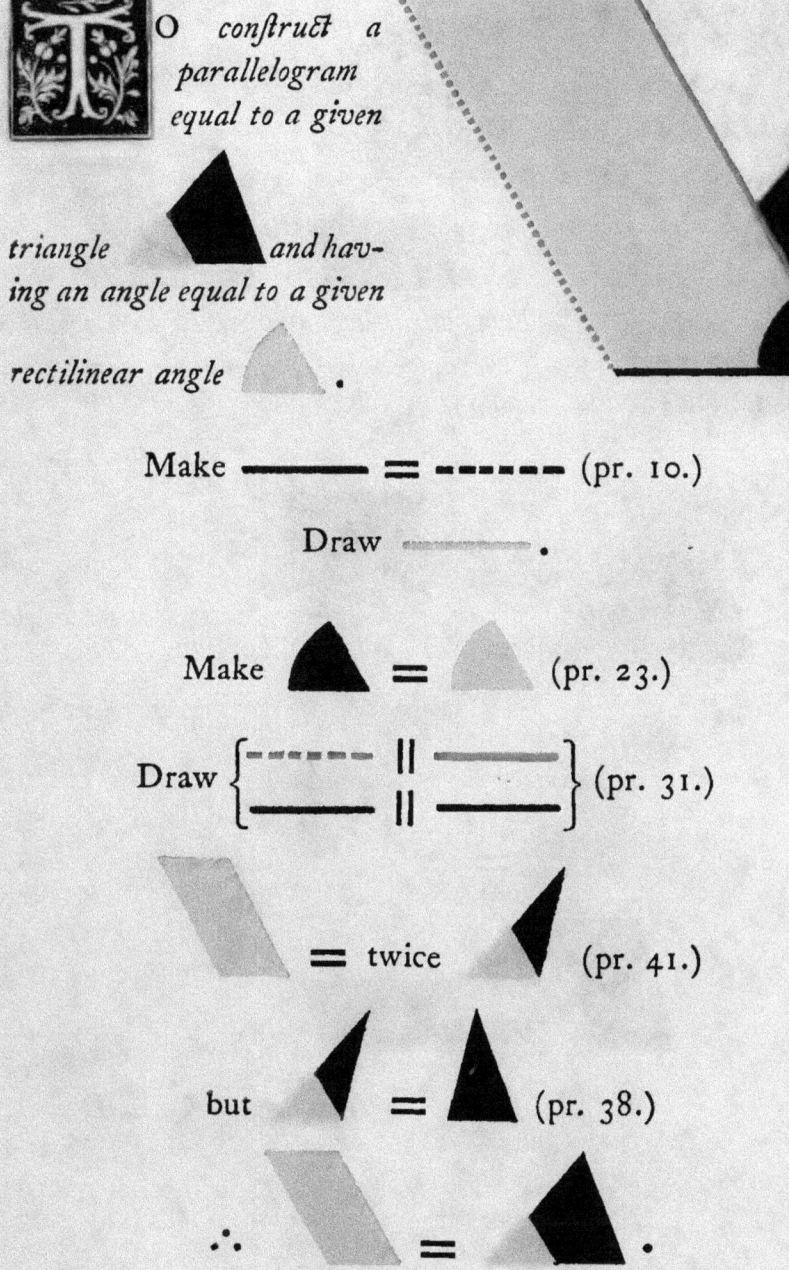

T O construct a parallelogram equal to a given triangle and having an angle equal to a given rectilinear angle .

Make ———— = —————— (pr. 10.)

Draw ———————— .

Make ▲ = ▲ (pr. 23.)

Draw { ———— || ———— } (pr. 31.)

= twice ◀ (pr. 41.)

but ◀ = ▲ (pr. 38.)

∴ ▱ = ◣ .

Q. E. D.

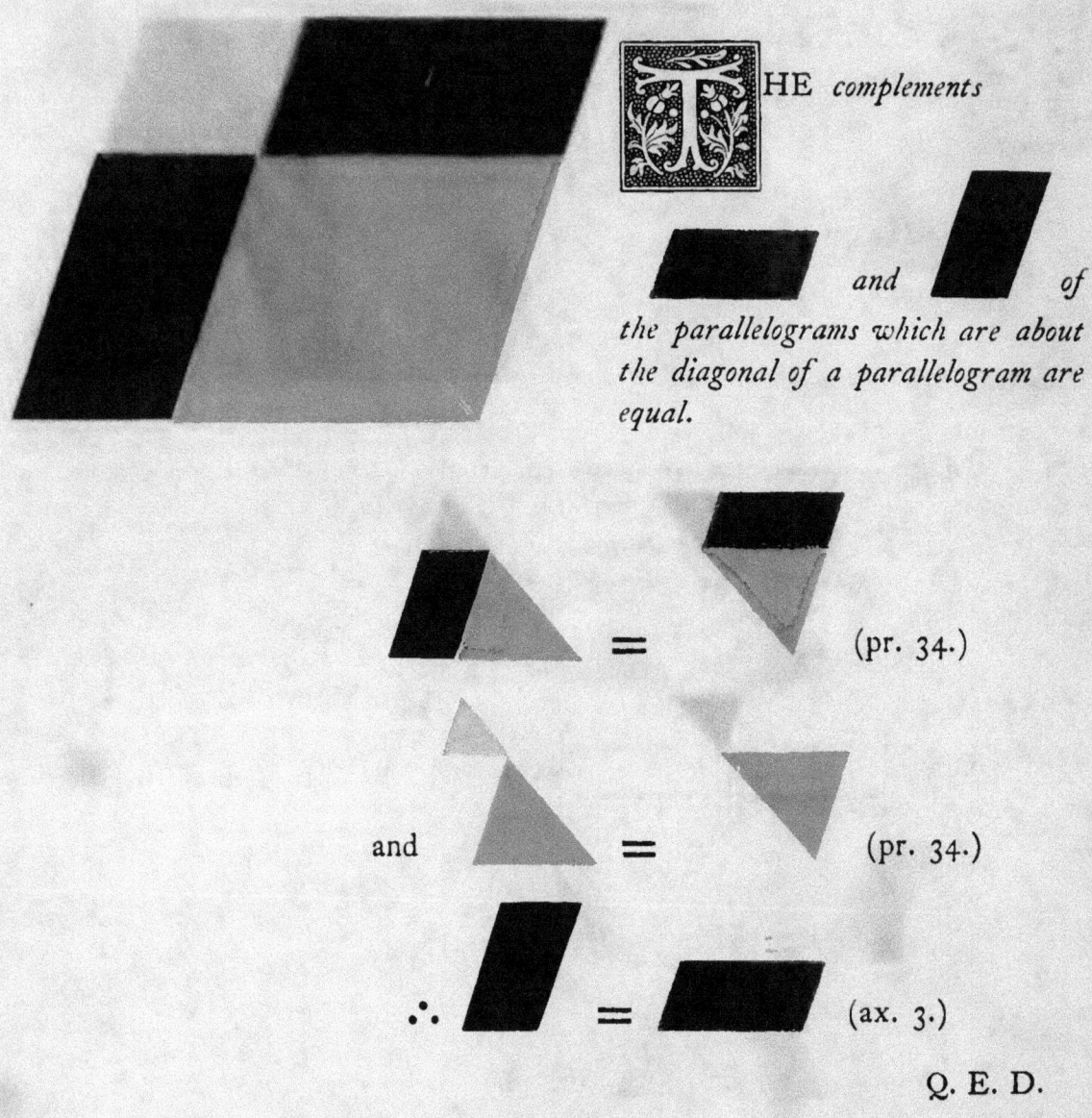

THE *complements* and of the parallelograms which are about the diagonal of a parallelogram are equal.

= (pr. 34.)

and = (pr. 34.)

∴ = (ax. 3.)

Q. E. D.

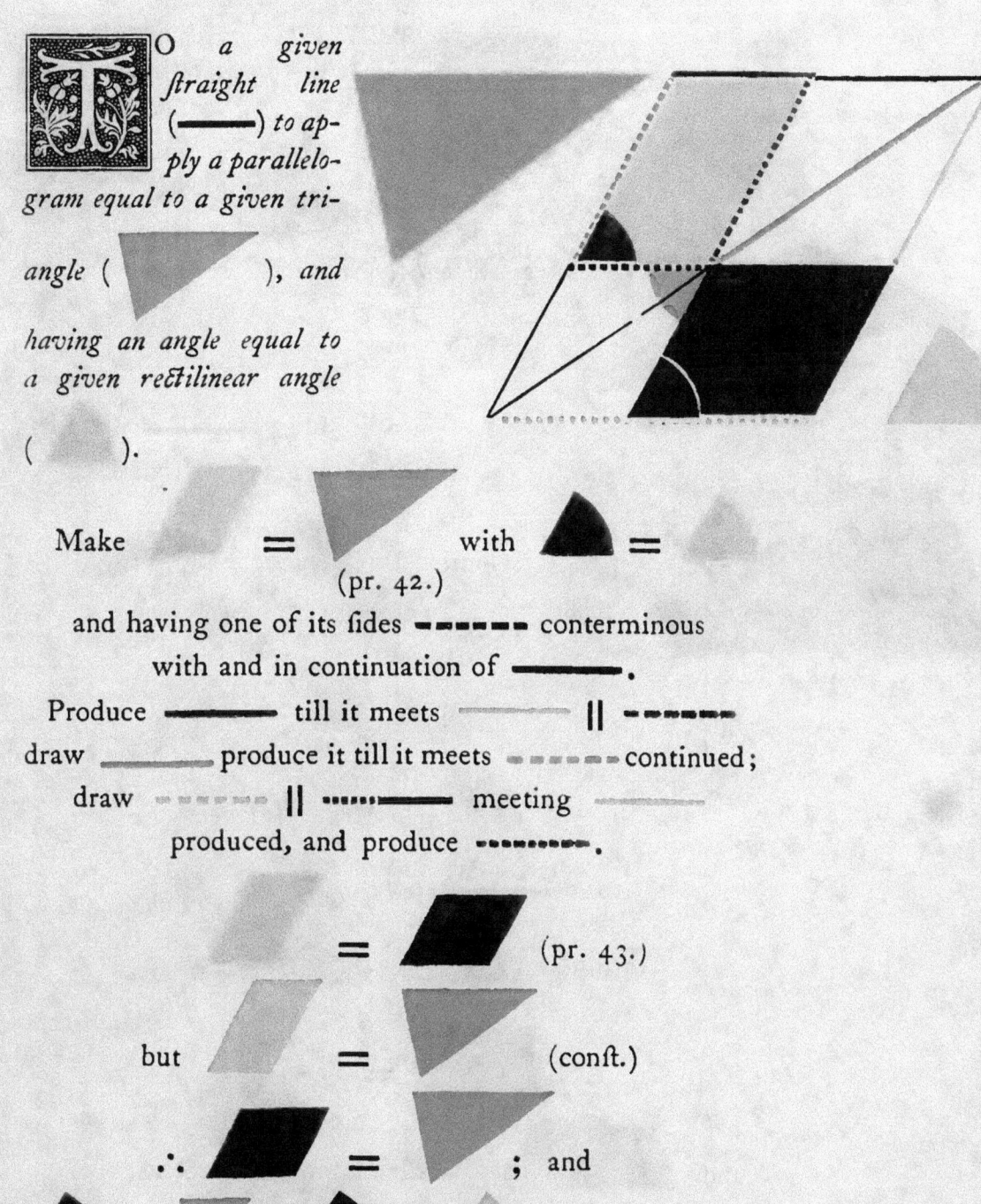

O a given ſtraight line (———) to apply a parallelogram equal to a given triangle ( ), and having an angle equal to a given rectilinear angle ( ).

Make   =   with   =
(pr. 42.)
and having one of its ſides ▬▬▬▬ conterminous with and in continuation of ———.
Produce ——— till it meets ——— ‖ ▬▬▬▬
draw ——— produce it till it meets ▬ ▬ ▬ continued;
draw ▬▬▬ ‖ ▬▬▬ meeting ——— produced, and produce ▬▬▬▬.

=   (pr. 43.)

but   =   (conſt.)

∴   =   ; and

=   =   =   (pr. 19. and conſt.)
Q. E. D.

O *conſtruct a parallelogram equal to a given rectilinear figure*

(   ) *and having an*

angle equal to a given rectilinear angle

(   ).

Draw ——— and ——— dividing the rectilinear figure into triangles.

Conſtruct    =

having    =    (pr. 42.)

to ——— apply    =

having    =    (pr. 44.)

to ——— apply    =

having    =    (pr. 44.)

∴    =

and    is a parallelogram. (prs. 29, 14, 30.)

having    =    .

Q. E. D.

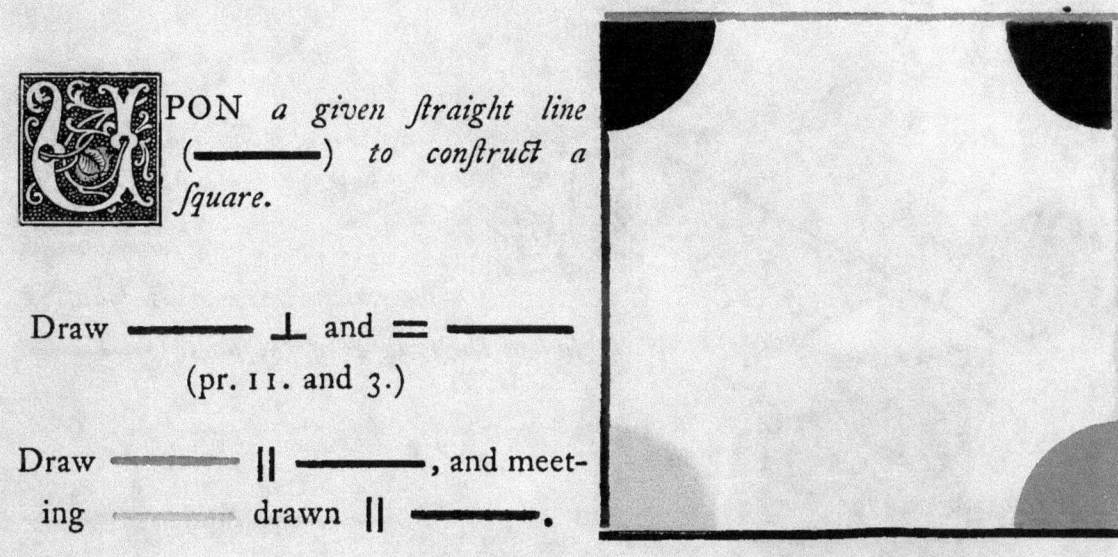

PON *a given straight line* (————) *to construct a square.*

Draw ———— ⊥ and ═ ————
(pr. 11. and 3.)

Draw ———— ‖ ————, and meeting ———— drawn ‖ ————.

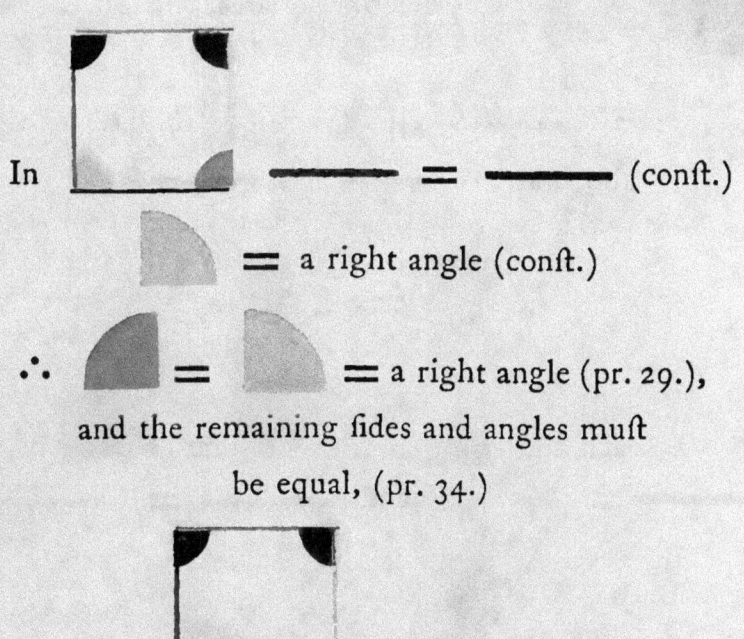

In ———— ═ ———— (conſt.)

═ a right angle (conſt.)

∴ ═ ═ a right angle (pr. 29.),

and the remaining ſides and angles muſt

be equal, (pr. 34.)

and ∴ is a ſquare. (def. 27.)

Q. E. D.

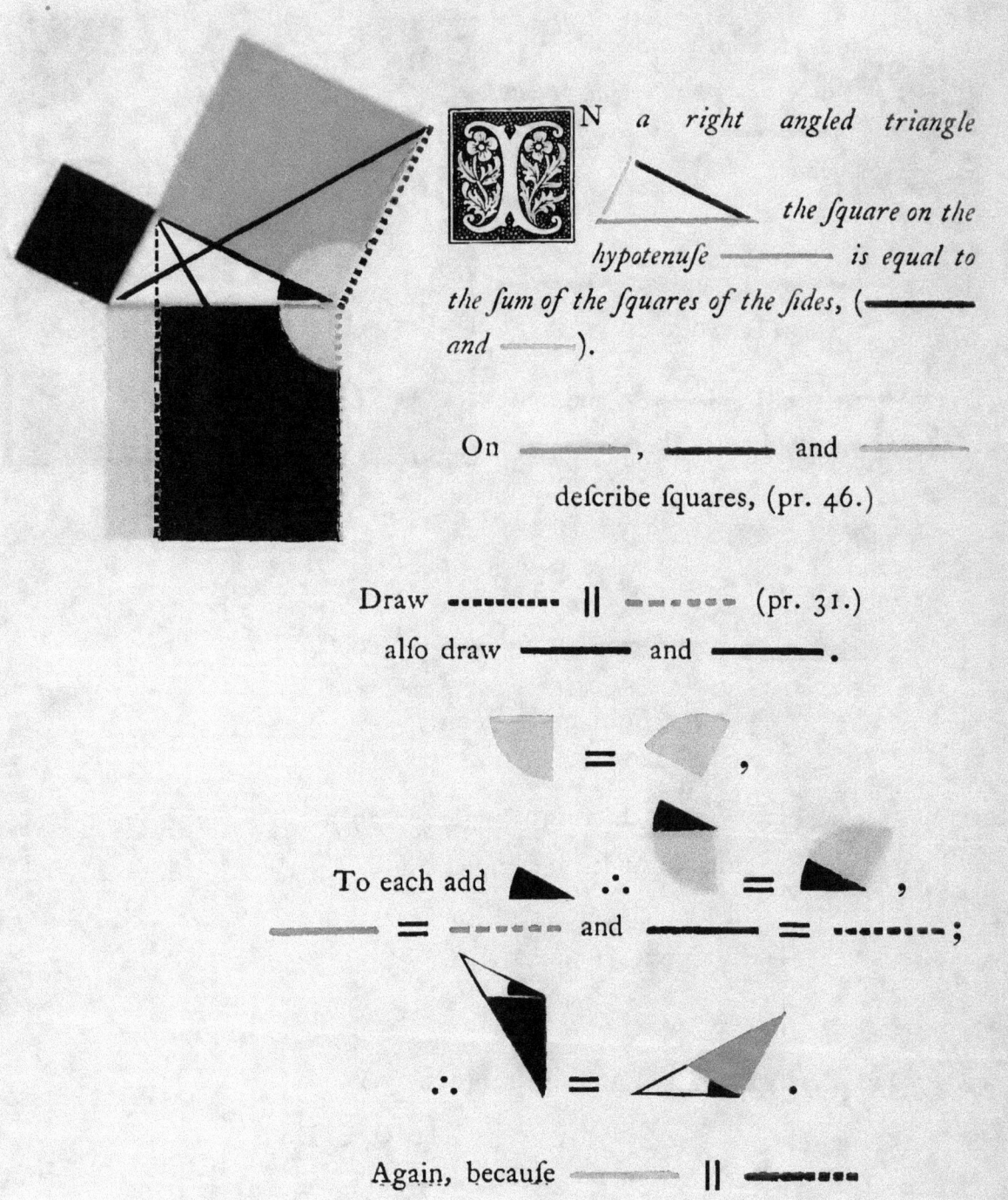

N a right angled triangle the square on the hypotenufe ———— is equal to the fum of the fquares of the fides, (———— and ————).

On ————, ———— and ———— defcribe fquares, (pr. 46.)

Draw ••••••••• || ‑ ‑ ‑ ‑ (pr. 31.)
alfo draw ———— and ————.

$=$,

To each add ∴ = ,
= and = ;

∴ = .

Again, becaufe ———— || ————

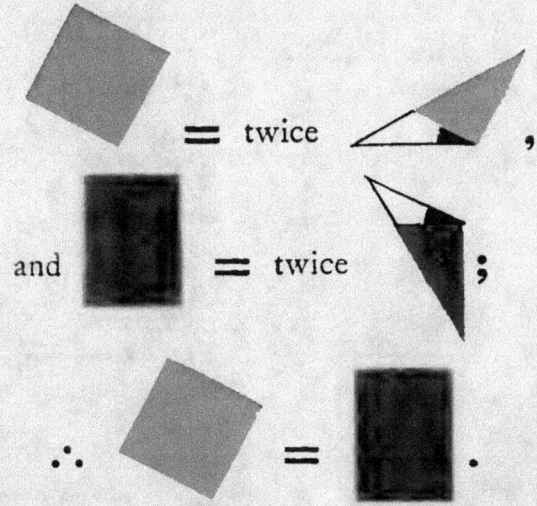

In the fame manner it may be fhown

that ■ = ▮ ;

hence ◆ = ■ .

Q. E. D.

H

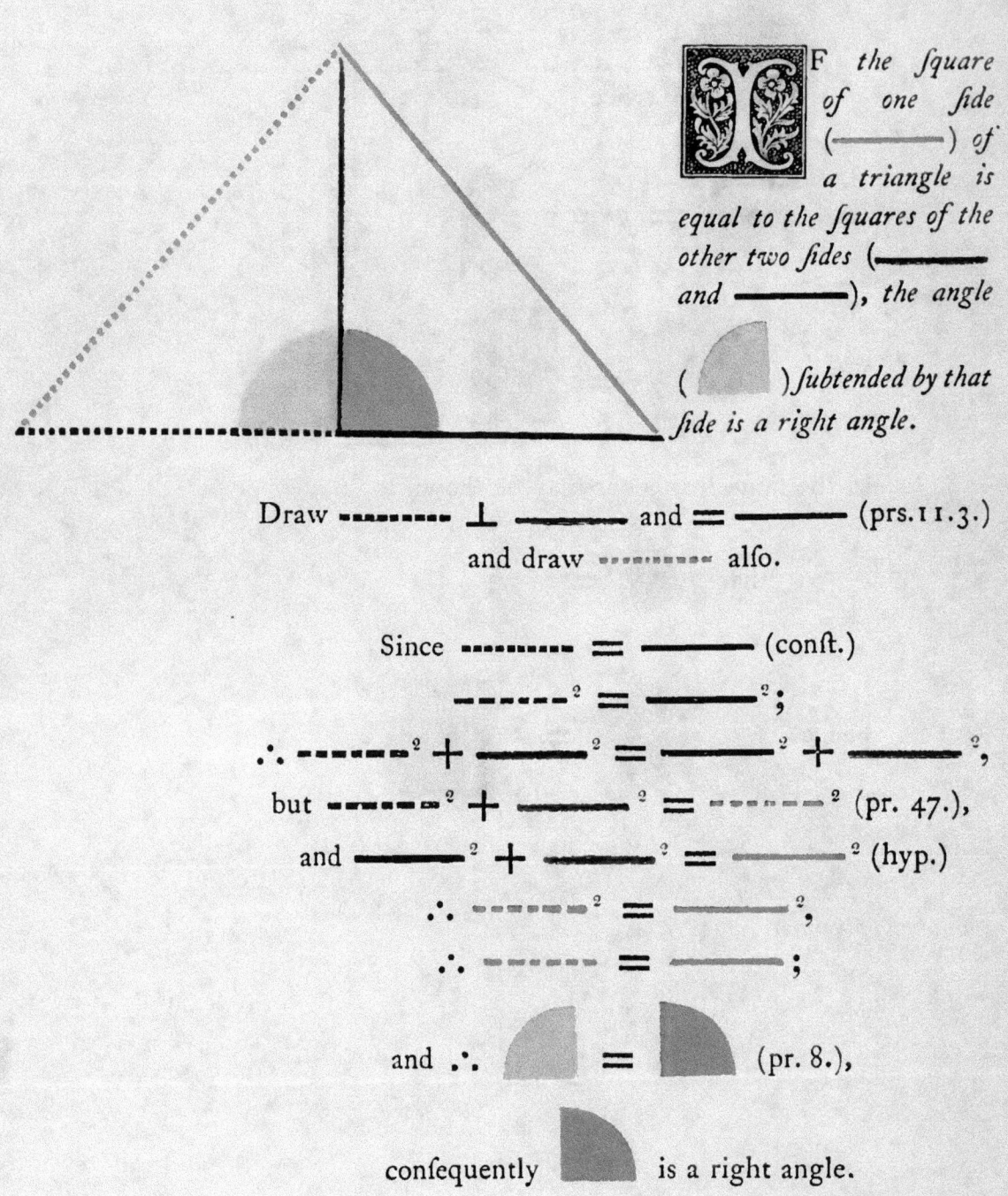

IF the square of one side (———) of a triangle is equal to the squares of the other two sides (——— and ———), the angle (◳) subtended by that side is a right angle.

Draw ▪▪▪▪▪▪▪ ⊥ ——— and ═ ——— (prs.11.3.) and draw ▪▪▪▪▪▪ also.

Since ▪▪▪▪▪▪▪ ═ ——— (conft.)

▪▪▪▪▪▪▪² ═ ———²;

∴ ▪▪▪▪▪▪▪² + ——— ² ═ ———² + ———²,

but ▪▪▪▪▪▪▪² + ———² ═ ▪▪▪▪▪▪² (pr. 47.),

and ———² + ———² ═ ———² (hyp.)

∴ ▪▪▪▪▪▪² ═ ———²,

∴ ▪▪▪▪▪▪ ═ ———;

and ∴ ◳ ═ ◳ (pr. 8.),

consequently ◳ is a right angle.

Q. E. D.

BOOK II.

DEFINITION I.

 RECTANGLE or a right angled parallelogram is said to be contained by any two of its adjacent or conterminous sides.

Thus: the right angled parallelogram ▬ is said to be contained by the sides ▬▬▬ and ▬▬▬ ; or it may be briefly designated by

▬▬▬ · ▬▬▬ .

If the adjacent sides are equal; i. e. ▬▬▬ = ▬▬▬ , then ▬▬▬ · ▬▬▬ which is the expression for the rectangle under ▬▬▬ and ▬▬▬ is a square, and

is equal to $\begin{cases} \text{▬▬▬} \cdot \text{▬▬▬ or ▬▬▬}^2 \\ \text{▬▬▬} \cdot \text{▬▬▬ or ▬▬▬}^2 \end{cases}$

DEFINITION II.

I N a parallelogram, the figure compofed of one of the parallelograms about the diagonal, together with the two complements, is called a *Gnomon*.

Thus 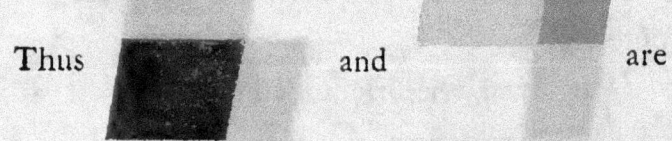 and   are

called Gnomons.

 HE *rectangle contained by two straight lines, one of which is divided into any number of parts,*

is equal to the sum of the rectangles contained by the undivided line, and the several parts of the divided line.

Draw ▬▬ ⊥ ▬ ▬▬ and ＝ ▬▬ (prs. 2. 3. B. 1.);
complete the parallelograms, that is to say,

Draw { ... } (pr. 31. B. 1.)

∴ Q. E. D.

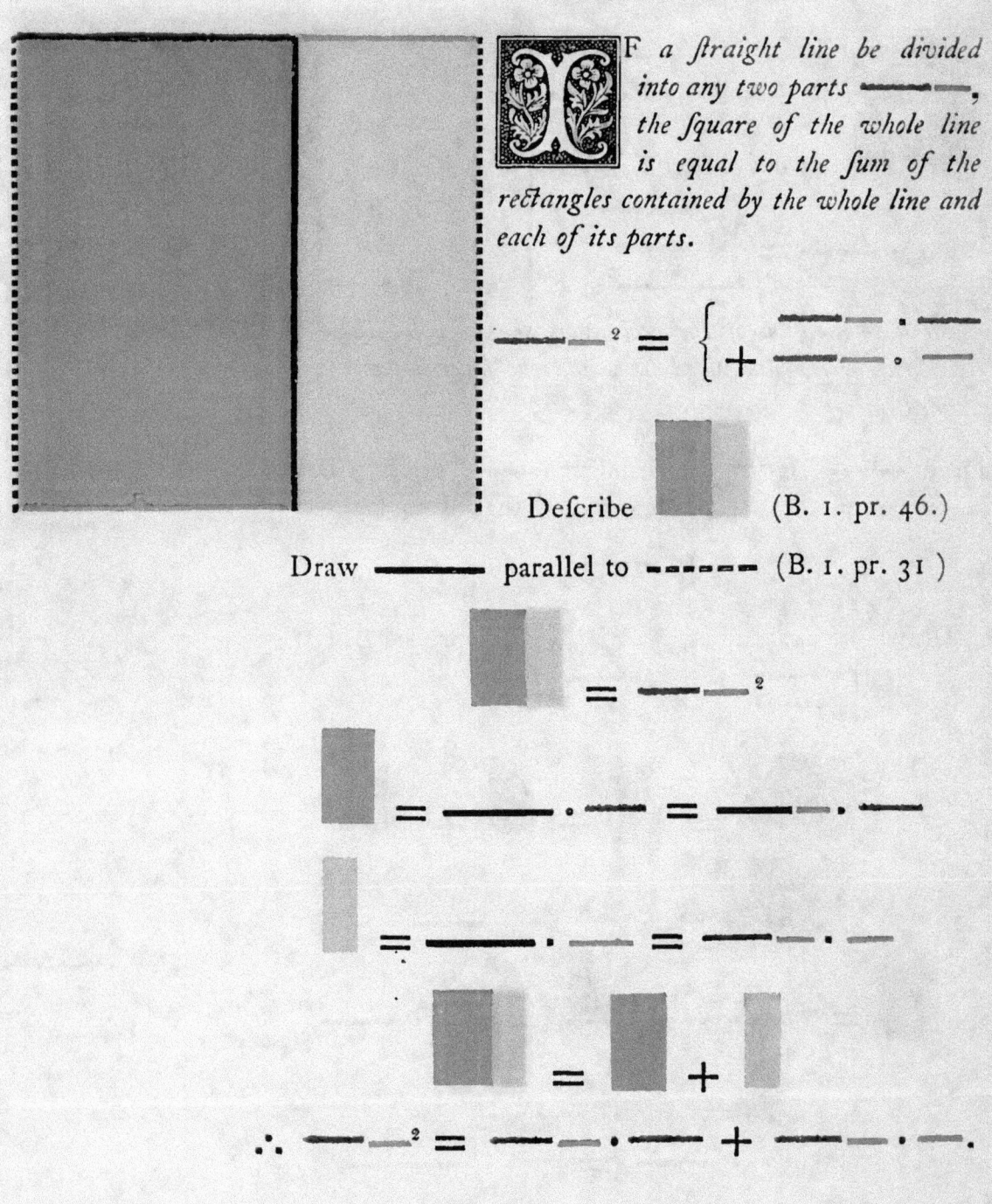

F a ſtraight line be divided into any two parts ———— ——, the ſquare of the whole line is equal to the ſum of the rectangles contained by the whole line and each of its parts.

Deſcribe   (B. 1. pr. 46.)

Draw ———— parallel to ------- (B. 1. pr. 31)

Q. E. D.

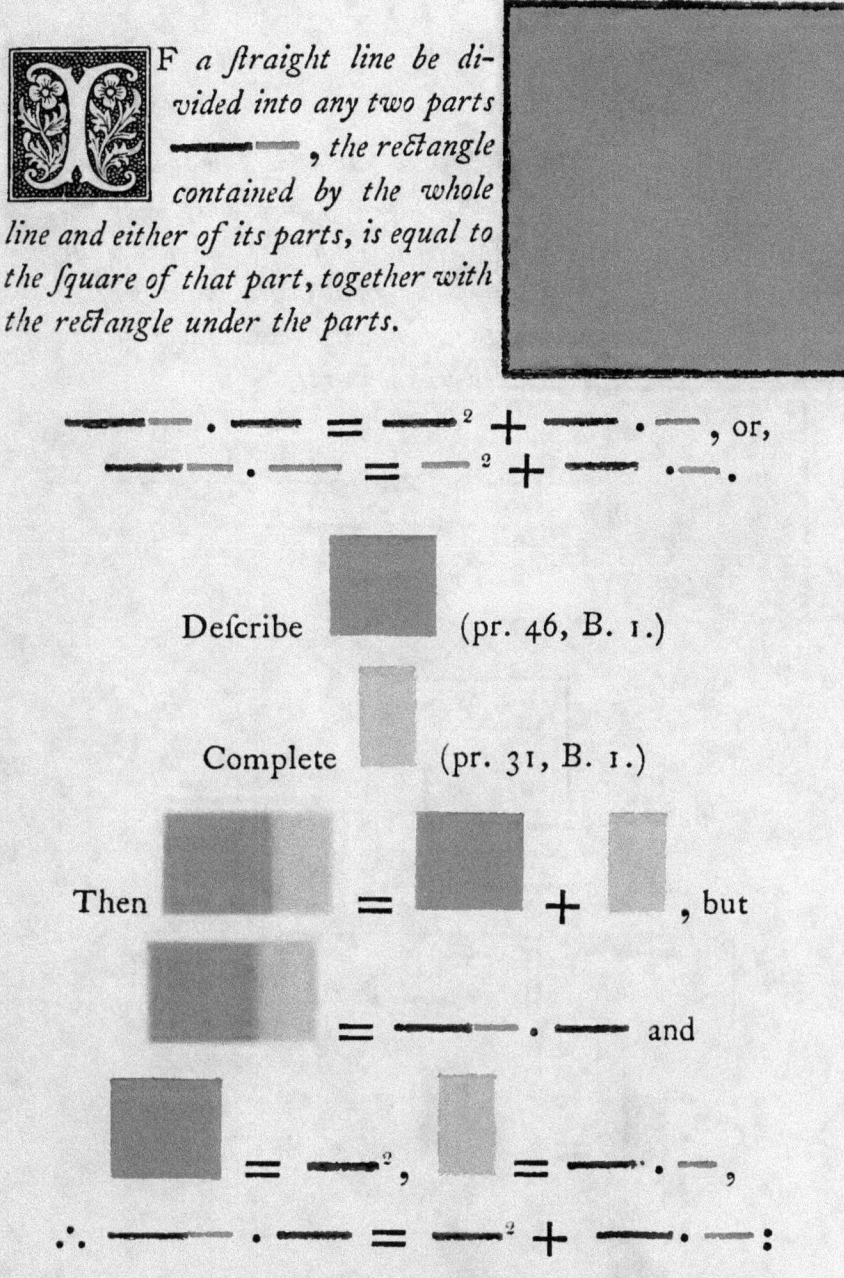

F *a straight line be di-
vided into any two parts
———— ——, the rectangle
contained by the whole
line and either of its parts, is equal to
the square of that part, together with
the rectangle under the parts.*

———— · —— = ————² + —— · ——, or,

———— · —— = ————² + —— · ——.

Deſcribe ▢ (pr. 46, B. 1.)

Complete ▯ (pr. 31, B. 1.)

Then ▢ = ▢ + ▯, but

▢ = ———— · —— and

▢ = ————², ▯ = —— · ——,

∴ ———— · —— = ————² + —— · —— :

In a ſimilar manner it may be readily ſhown

that ———— · —— = ——² + ———— · ——.

Q. E. D

F *a straight line be divided into any two parts* ▬▬ ▬, *the square of the whole line is equal to the squares of the parts, together with twice the rectangle contained by the parts.*

$$\text{▬▬}^{2} = \text{▬▬}^{2} + \text{▬▬}^{2} +$$
$$\text{twice } \text{▬▬} \cdot \text{▬▬}.$$

Defcribe ▢ (pr. 46, B. 1.)

draw ▬▬▬ ▪▪▪▪ (poft. 1.),

and $\left\{ \begin{array}{c} \text{▬▬▪▪} \\ \text{▪▪▬▬} \end{array} \middle\| \begin{array}{c} \text{▬▬▬} \\ \text{▬▬▬} \end{array} \right\}$ (pr. 31, B. 1.)

◣ = ◤ (pr. 5, B. 1.),

◣ = ◤ (pr. 29, B. 1.)

∴ ◤ = ◤

∴ by (prs. 6, 29, 34. B. 1.) ◨ is a square $=$ ▬².

For the fame reafons ◨ is a square $=$ ▬²,

▬ $=$ ▮ $=$ ▬ . ▬ (pr. 43, b. 1.)

but ◲ $=$ ◨ $+$ ▬ $+$ ▮ $+$ ◱,

∴ ▬▬² $=$ ▬² $+$ ▬² $+$

twice ▬ . ▬.

Q. E. D.

I

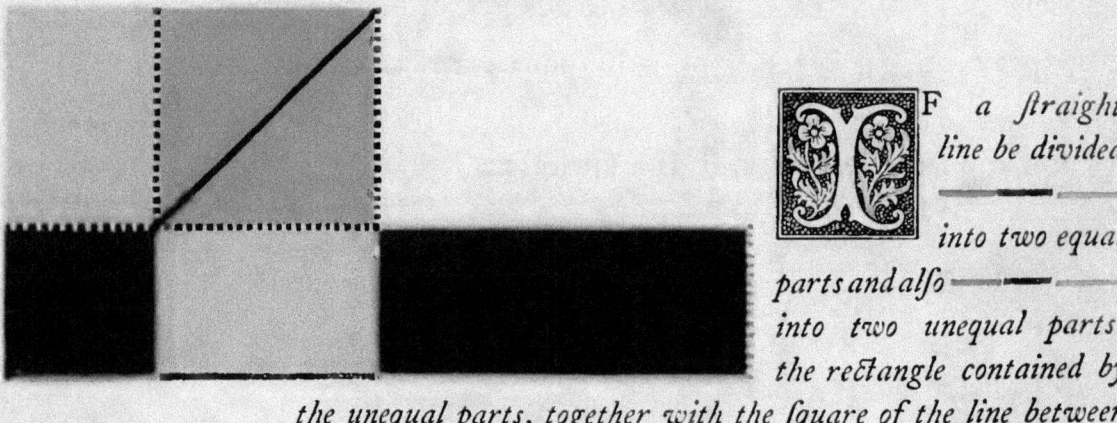

F a *ſtraight line be divided* into two equal parts and alſo into two unequal parts, the rectangle contained by the unequal parts, together with the ſquare of the line between the points of ſection, is equal to the ſquare of half that line

$$ \underline{\quad} \cdot \underline{\quad} + \underline{\quad}^2 = \underline{\quad}^2 = \underline{\quad}^2, $$

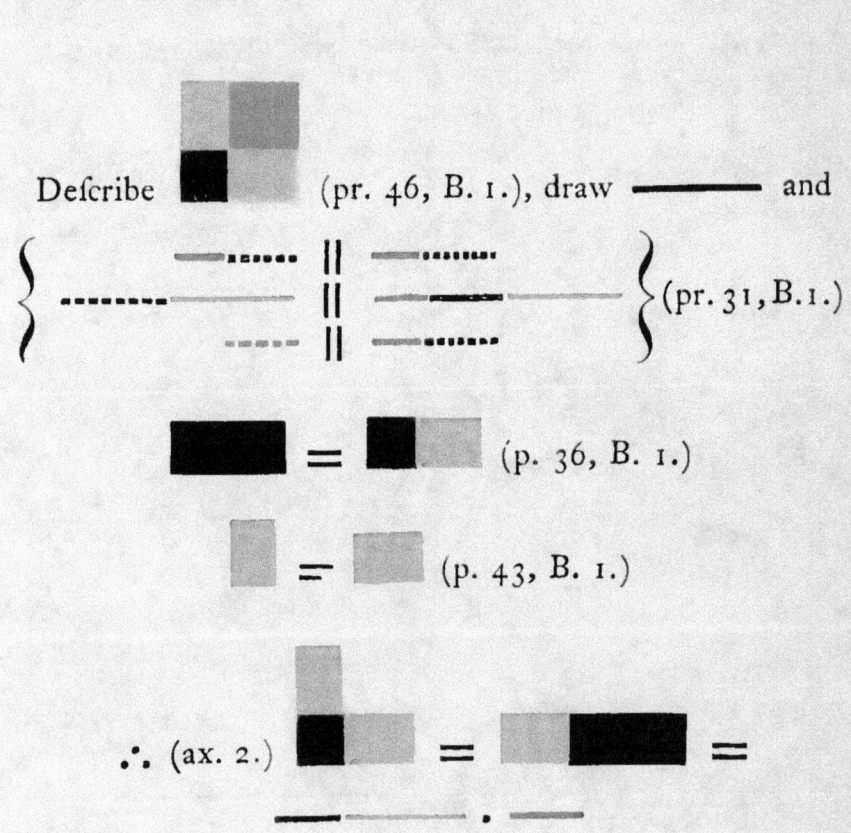

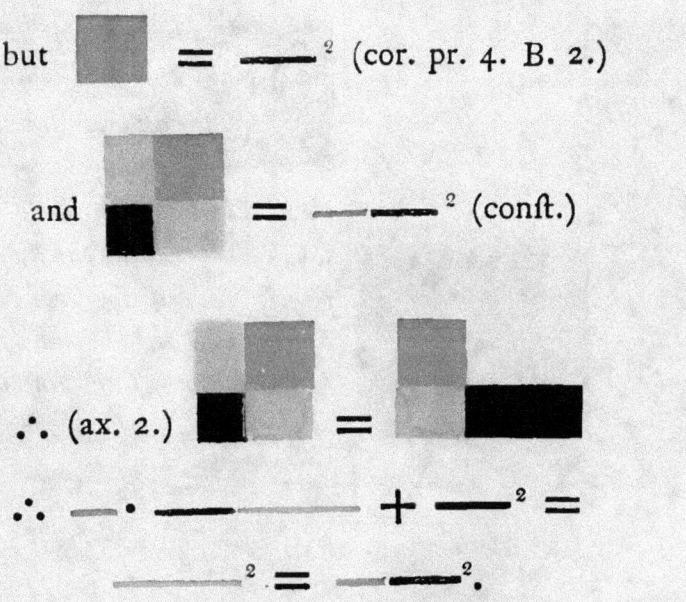

but ▪ = ——² (cor. pr. 4. B. 2.)

and ▪ = —— ——² (conſt.)

∴ (ax. 2.) ▪ = ▪

∴ —— • —————— + ——² =

—————² = —— ——².

Q. E. D.

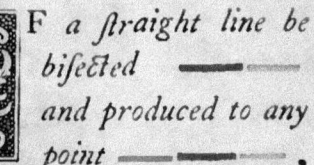 F *a straight line be bisected* ▬ *and produced to any point* ▬▬ ; *the rectangle contained by the whole line so increased, and the part produced, together with the square of half the line, is equal to the square of the line made up of the half, and the produced part.*

$$ \underline{\quad} \cdot \underline{\quad} + \underline{\quad}^{2} = \underline{\quad}^{2}. $$

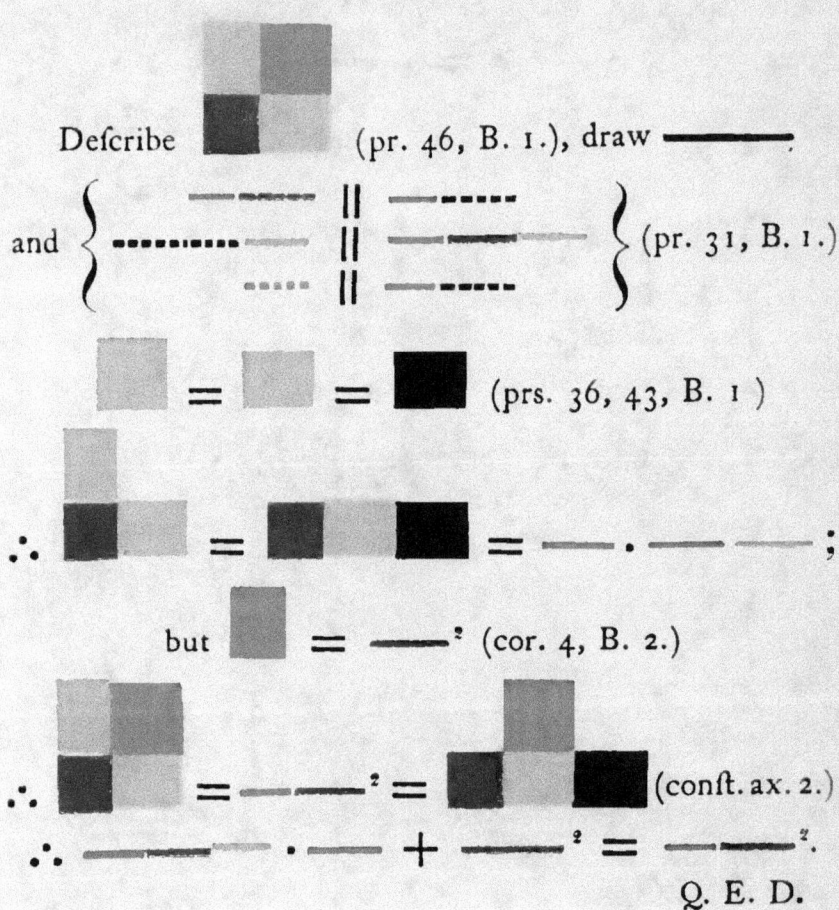

Describe ▨ (pr. 46, B. 1.), draw ▬▬▬ .

and $\left\{ \begin{array}{c} \dots \ \| \ \dots \\ \dots \ \| \ \dots \\ \dots \ \| \ \dots \end{array} \right\}$ (pr. 31, B. 1.)

▨ $=$ ▨ $=$ ■ (prs. 36, 43, B. 1)

∴ ▨ $=$ ■ $=$ ▬·▬ ;

but ▨ $=$ ▬² (cor. 4, B. 2.)

∴ ▨ $=$ ▬² $=$ ■ ▨ (const. ax. 2.)

∴ ▬ · ▬ $+$ ▬² $=$ ▬² .

Q. E. D.

60

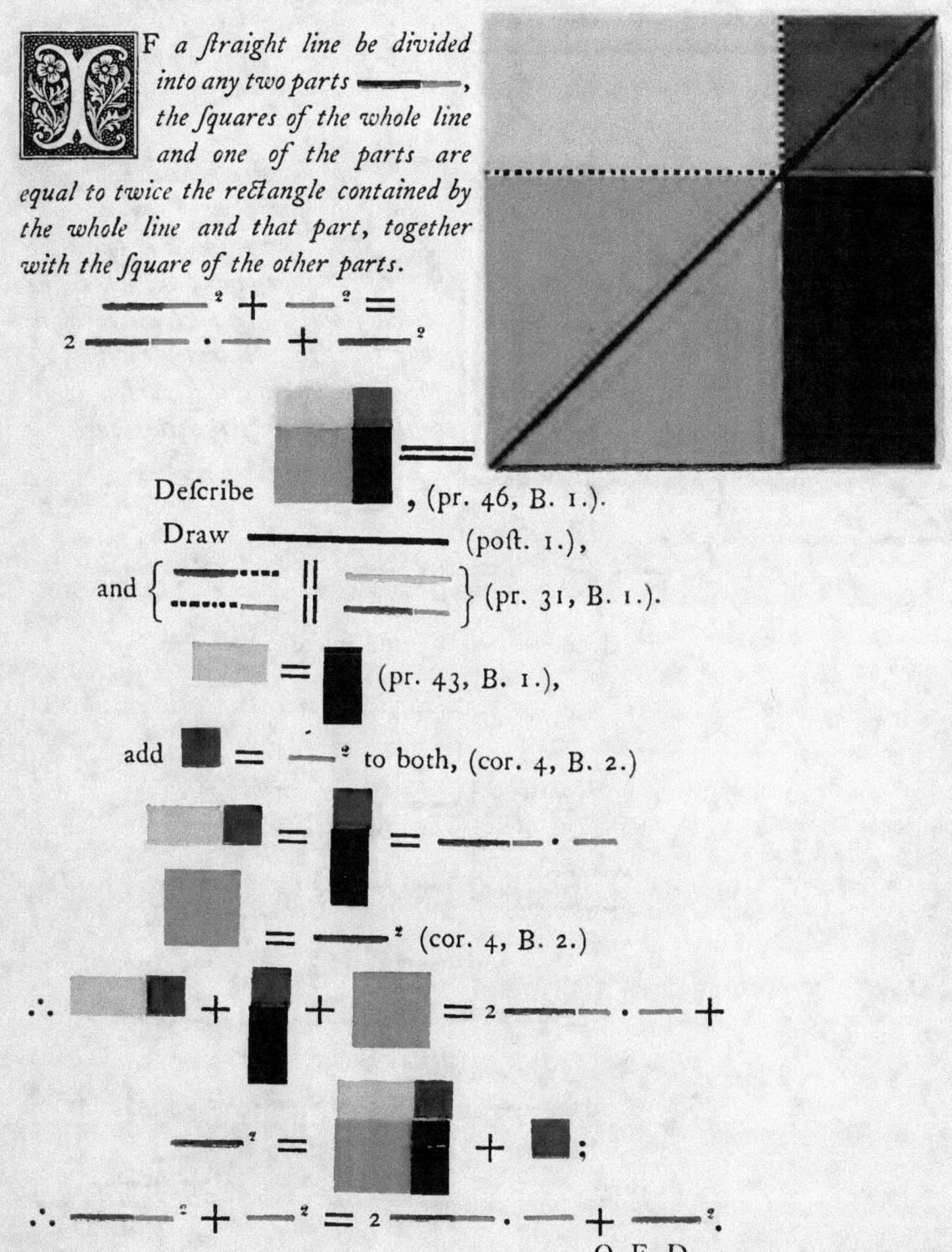

F a ſtraight line be divided into any two parts ——, the ſquares of the whole line and one of the parts are equal to twice the rectangle contained by the whole line and that part, together with the ſquare of the other parts.

Deſcribe ▦ , (pr. 46, B. 1.).

Draw ———— (poſt. 1.),

and ⎰ ⎱ (pr. 31, B. 1.).

▭ = ▮ (pr. 43, B. 1.),

add ■ = ——² to both, (cor. 4, B. 2.)

▭ = ▮ = 2 ————·——

▦ = ——² (cor. 4, B. 2.)

∴ ▭ + ▮ + ▦ = 2 ———— ·—— +

——² = ▦ + ■ ;

∴ ——² + ——² = 2 ———— ·—— + ——².

Q. E. D.

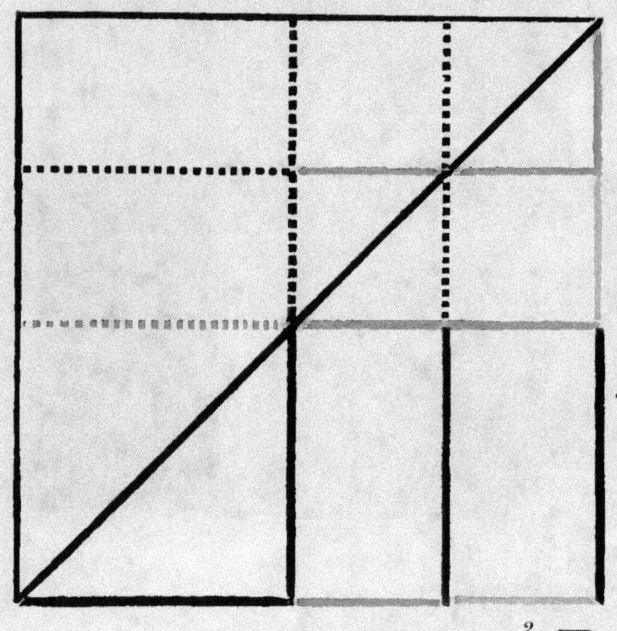

F *a straight line be divided into any two parts* ———, *the square of the sum of the whole line and any one of its parts, is equal to four times the rectangle contained by the whole line, and that part together with the square of the other part.*

$$\text{———}^2 = 4 \cdot \text{———} \cdot \text{———} + \text{———}^2,$$

Produce ——— and make ——— = ———

Construct  (pr. 46, B. 1.);

draw ———————— ,

$$\left.\begin{matrix}\text{———}\\ \text{———}\end{matrix}\right\} \parallel \text{———} \left.\begin{matrix}\\ \\ \\ \\\end{matrix}\right\}$$

$$\left.\begin{matrix}\text{———}\\ \text{———}\end{matrix}\right\} \parallel \text{———} \quad \text{(pr. 31, B. 1.)}$$

$$\text{———}^2 = \text{———}^2 + \text{———}^2 + 2 \cdot \text{———} \cdot \text{———}$$
$$\text{(pr. 4, B. 11.)}$$

but $\text{———}^2 + \text{———}^2 = 2 \cdot \text{———} \cdot \text{———} + \text{———}^2$
$$\text{(pr. 7, B. 11.)}$$

$$\therefore \; \text{———}^2 = 4 \cdot \text{———} \cdot \text{———} + \text{———}^2.$$

Q. E. D.

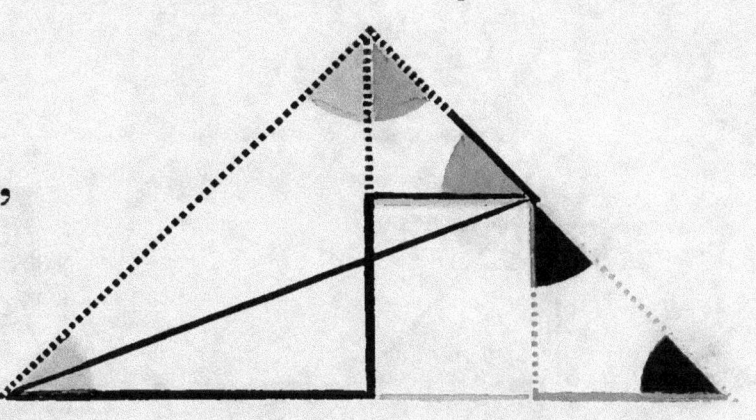

F *a ftraight line be divided into two equal parts* ━━ , *and alfo into two unequal parts* ━━ , *the fquares of the unequal parts are together double the fquares of half the line, and of the part between the points of fection.*

$$\text{━━}^2 + \text{━}^2 = 2\,\text{━━}^2 + 2\,\text{━}^2.$$

Make ━━ ⊥ and ═ ━━ or ━━ ,

Draw ⋯⋯ and ⋯⋯ ,

━━ ‖ ━━ , ━ ‖ ━━ , and draw ━━ .

▲ = ◢ (pr. 5, B. 1.) ═ half a right angle.
(cor. pr. 32, B. 1.)

▲ = ◥ (pr. 5, B. 1.) ═ half a right angle.
(cor. pr. 32, B. 1.)

∴ ◣◢ ═ a right angle.

▲ = ◢ = ▲ = ◥

(prs. 5, 29, B. 1.).

hence ━━ = ━━ , ━━ = ━━ = ━━

(prs. 6, 34, B. 1.)

$$\text{━━}^2 = \begin{cases} \text{━━}^2 + \text{━}^2, \text{ or } + \text{━}^2 \\ = \end{cases} \begin{cases} \text{⋯}^2 = 2\,\text{━━}^2 \end{cases}$$

(pr. 47, B. 1.)

$$\text{⋯}^2 = 2\,\text{━}^2$$

∴ $\text{━━}^2 + \text{━}^2 = 2\,\text{━━}^2 + 2\,\text{━}^2.$

Q. E. D.

F a *ftraight line be bi-fected and pro-duced to any point* ———, *the fquares of the whole produced line, and of the produced part, are toge-ther double of the fquares of the half line, and of the line made up of the half and pro-duced part.*

$$\underline{\quad\quad}^2 + \underline{\quad}^2 = 2\,\underline{\quad}^2 + 2\,\underline{\quad}^{\;2}.$$

Make ▬▬ ⊥ and = to —— or ——,

draw ▬▬••• and ▬▬---- ,

and $\left\{ \begin{array}{c} \text{▬▬•••} \\ \text{•••▬▬} \end{array} \right\|$ ‖ $\begin{array}{c} \text{▬▬} \\ \text{▬▬} \end{array}$ $\Big\}$ (pr. 31, B. 1.);

draw ▬▬ alfo.

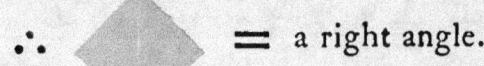

(pr. 5, B. 1.) = half a right angle.
(cor. pr. 32, B. 1.)

(pr. 5, B. 1.) = half a right angle
(cor. pr. 32, B. 1.)

∴ = a right angle.

half a right angle (prs. 5, 32, 29, 34, B. 1.),

and $\rule{1cm}{0.5pt}$ = $\rule{1cm}{0.5pt}$, $\rule{1cm}{0.5pt}$ = $\rule{1cm}{0.5pt}$ =

$\rule{1cm}{0.5pt}$, (prs. 6, 34, B. 1.). Hence by (pr. 47, B. 1.)

$$\rule{1cm}{0.5pt}^2 = \begin{cases} \rule{2cm}{0.5pt}^2 + \rule{1cm}{0.5pt}^2 \text{ or } \rule{1cm}{0.5pt}^2 \\ \begin{cases} + \rule{1cm}{0.5pt}^2 = 2 \rule{1cm}{0.5pt}^2 \\ + \rule{2cm}{0.5pt}^2 = 2 \rule{1cm}{0.5pt}^2 \end{cases} \end{cases}$$

$\therefore \rule{1cm}{0.5pt}^2 + \rule{1cm}{0.5pt}^2 = 2 \rule{1cm}{0.5pt}^2 + 2 \rule{1cm}{0.5pt}^2.$

Q. E. D.

K

65

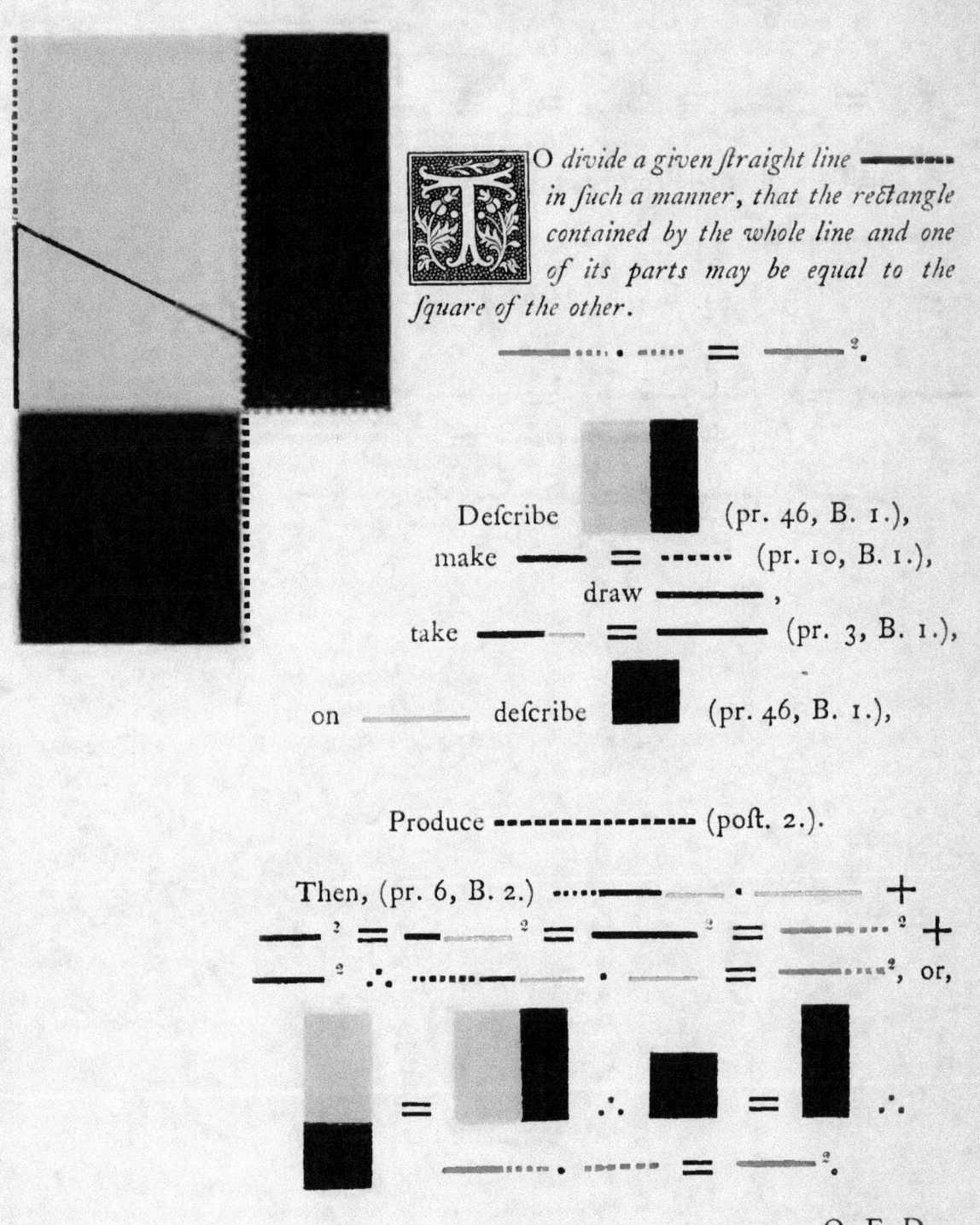

T O divide a given *straight line* ▬▬ ▪▪▪▪ in *such a manner, that the rectangle contained by the whole line and one of its parts may be equal to the square of the other.*

▬▬ ▪▪▪ ▪ ▪▪▪ = ▬▬².

Defcribe ▮▮ (pr. 46, B. 1.),

make ▬▬ = ▪▪▪▪▪ (pr. 10, B. 1.),

draw ▬▬ ,

take ▬▬ = ▬▬▬ (pr. 3, B. 1.),

on ▬▬ defcribe ■ (pr. 46, B. 1.),

Produce ▪▪▪▪▪▪▪▪▪ (poft. 2.).

Then, (pr. 6, B. 2.) ▪▪▪ ▬ ▬ ▪ ▬ +

▬▬² = ▬ ▬▬² = ▬▬² = ▬▬² +

▬▬² ∴ ▪▪▪▪▪ ▪ ▬ = ▬▬², or,

■ = ■■ ∴ ■ = ■ ∴

▬▬ ▪ ▪▪▪ = ▬▬².

Q. E. D.

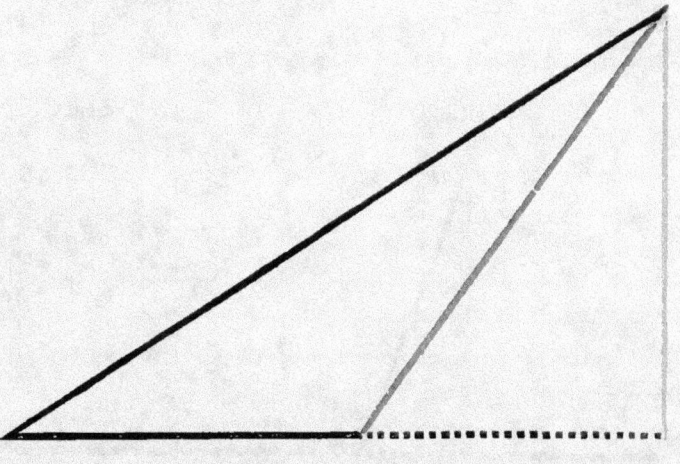

N *any obtuse angled triangle, the square of the side subtending the obtuse angle exceeds the sum of the squares of the sides containing the obtuse angle, by twice the rectangle contained by either of these sides and the produced parts of the same from the obtuse angle to the perpendicular let fall on it from the opposite acute angle.*

$$\rule{2cm}{1pt}^2 \;\sqsubset\; \rule{1.5cm}{1pt}^2 + \rule{1.5cm}{1pt}^2 \text{ by}$$

$$2\,\rule{1.5cm}{1pt}\cdot\rule{1.5cm}{0.5pt}.$$

By pr. 4, B. 2.

$$\rule{2cm}{1pt}^2 = \rule{1.5cm}{1pt}^2 + \rule{1.5cm}{0.5pt}^2 + 2\,\rule{1.5cm}{1pt}\cdot\rule{1cm}{0.5pt}:$$

add $\rule{1.5cm}{1pt}^2$ to both

$$\rule{1.5cm}{0.5pt}^2 + \rule{1.5cm}{1pt}^2 = \rule{1.5cm}{1pt}^2 \quad (\text{pr. 47, B. 1.})$$

$$= 2\cdot\rule{1.5cm}{1pt}\cdot\rule{1cm}{0.5pt} + \rule{1.5cm}{1pt}^2 + \left\{ \begin{matrix} \rule{1.5cm}{0.5pt}^2 \\ \rule{1.5cm}{1pt}^2 \end{matrix} \right\} \text{ or}$$

$$+ \rule{1.5cm}{1pt}^2 \quad (\text{pr. 47, B. 1.}). \quad \text{Therefore,}$$

$$\rule{1.5cm}{1pt}^2 = 2\cdot\rule{1.5cm}{1pt}\cdot\rule{1.5cm}{0.5pt} + \rule{1.5cm}{1pt}^2 +$$

$$\rule{1.5cm}{1pt}^2: \text{ hence } \rule{1.5cm}{1pt}^2 \sqsubset \rule{1.5cm}{1pt}^2 + \rule{1.5cm}{1pt}^2$$

$$\text{by } 2\cdot\rule{1.5cm}{1pt}\cdot\rule{1.5cm}{0.5pt}.$$

Q. E. D.

FIRST.   SECOND.

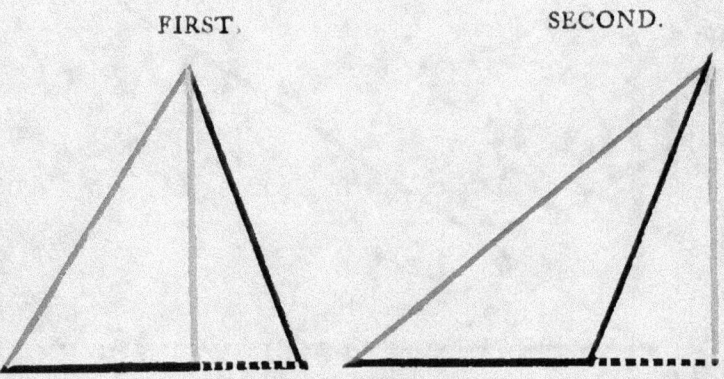

N any tri-
angle, the
*square of the
side subtend-*
*ing an acute angle, is
less than the sum of the
squares of the sides con-*
*taining that angle, by twice the rectangle contained by either
of these sides, and the part of it intercepted between the foot of
the perpendicular let fall on it from the opposite angle, and the
angular point of the acute angle.*

FIRST.

$$ \underline{\quad}^{2} \sqsupset \text{----}^{2} + \underline{\quad}^{2} \text{ by } 2 \cdot \text{------} \cdot \text{---} $$

SECOND.

$$ \underline{\quad}^{2} \sqsupset \underline{\quad}^{2} + \underline{\quad}^{2} \text{ by } 2 \cdot \text{---} \cdot \text{------} $$

First, suppose the perpendicular to fall within the
triangle, then (pr. 7, B. 2.)

$$ \text{------}^{2} + \underline{\quad}^{2} = 2 \cdot \text{------} \cdot \underline{\quad} + \text{------}^{2}, $$

add to each ────² then,

$$ \text{------}^{2} + \underline{\quad}^{2} + \underline{\quad}^{2} = 2 \cdot \text{------} \cdot \underline{\quad} $$
$$ + \text{------}^{2} + \underline{\quad}^{2} $$

∴ (pr. 47, B. 1.)

$$ \text{------}^{2} + \underline{\quad}^{2} = 2 \cdot \text{------} \cdot \underline{\quad} + \underline{\quad}^{2}, $$

and $\therefore$ ▬▬^{2} ⊐ ▬▬▬^{2} + ▬▬^{2} by

2 • ▬▬▬▬ • ▬▬ .

Next ſuppoſe the perpendicular to fall without the
triangle, then (pr. 7, B. 2.)

▬▬▬^{2} + ▬▬^{2} = 2 • ▬▬▬▬ • ▬▬ + ▬▬^{2},

add to each ▬▬^{2} then

▬▬▬^{2} + ▬▬^{2} + ▬▬^{2} = 2 • ▬▬▬ • ▬▬

+ ▬▬^{2} + ▬▬^{2} $\therefore$ (pr. 47, B. 1.),

▬▬^{2} + ▬▬^{2} = 2 • ▬▬▬▬ • ▬▬ + ▬▬^{2},

$\therefore$ ▬▬^{2} ⊐ ▬▬^{2} + ▬▬^{2} by 2 • ▬▬▬▬ • ▬▬ .

Q. E. D.

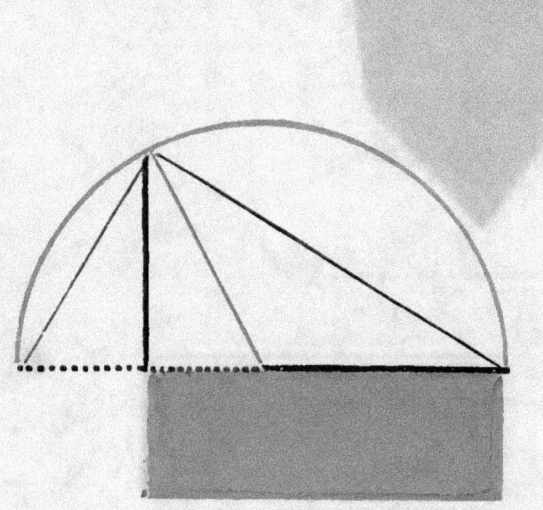

O *draw a right line of which the square shall be equal to a given rectilinear figure.*

To draw —————— *such that,*

$$ \text{——}^2 = $$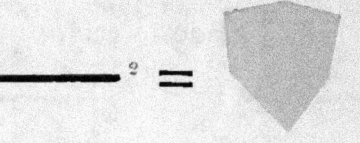

Make ▭ = ⬟ (pr. 45, B. 1.),

produce ·····——— until ·······= ——— ;

take ■·■·■·■ = ——————— (pr. 10, B. 1.),

Describe ⌒ (poſt. 3.),

and produce ——— to meet it : draw ———— .

$$ \text{——}^2 \text{ or } \text{——}^2 = \text{·······}\cdot\text{·······——} + \text{·······}^2 $$
(pr. 5, B. 2.),

but $$ \text{——}^2 = \text{——}^2 + \text{·······}^2 $$ (pr. 47, B. 1.);

∴ $$ \text{——}^2 + \text{·······}^2 = \text{·······}\cdot\text{·······——} + \text{·······}^2 , $$

∴ $$ \text{·—}^2 = \text{·······}\cdot\text{·······——} , \text{ and}$$

∴ $$ \text{——}^2 = \text{▭} = \text{⬟} $$

Q. E. D.

BOOK III.

DEFINITIONS.

I.

EQUAL circles are thofe whofe diameters are equal.

II.

A right line is said to touch a circle when it meets the circle, and being produced does not cut it.

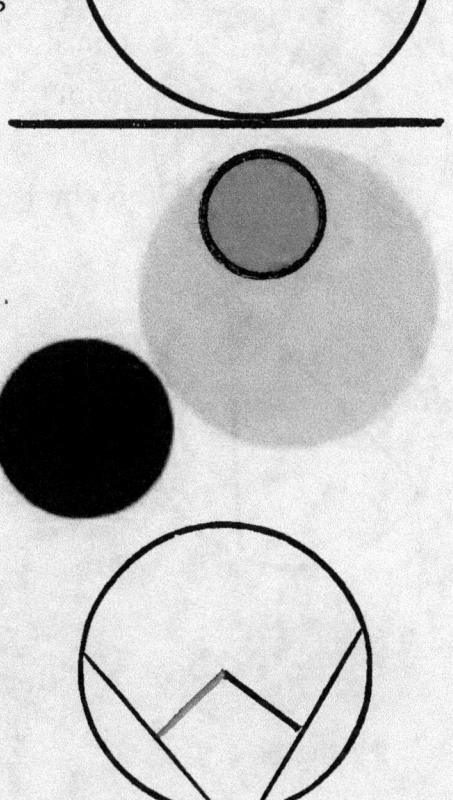

III.

Circles are faid to touch one another which meet but do not cut one another.

IV.

Right lines are faid to be equally diftant from the centre of a circle when the perpendiculars drawn to them from the centre are equal.

V.

And the ftraight line on which the greater perpendicular falls is faid to be farther from the centre.

VI.

A fegment of a circle is the figure contained by a ftraight line and the part of the circumference it cuts off.

VII.

An angle in a fegment is the angle contained by two ftraight lines drawn from any point in the circumference of the fegment to the extremities of the ftraight line which is the bafe of the fegment.

VIII.

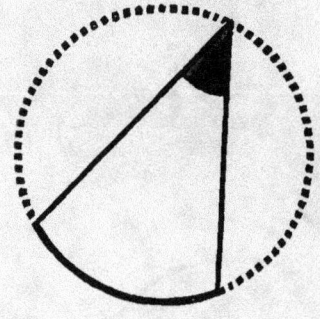

An angle is faid to ftand on the part of the circumference, or the arch, intercepted between the right lines that contain the angle.

IX.

A fector of a circle is the figure contained by two radii and the arch between them.

X.

Similar ſegments of circles are thoſe which contain equal angles.

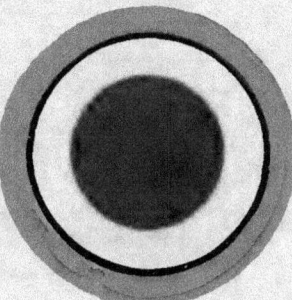

Circles which have the ſame centre are called *concentric circles*.

L

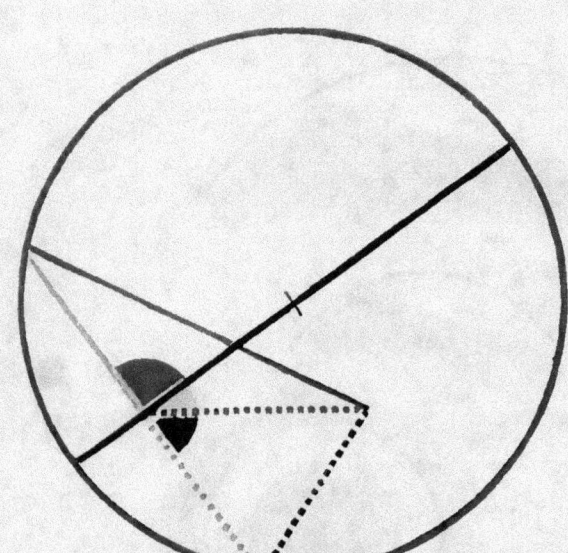

T O *find the centre of a given*

circle  .

Draw within the circle any ſtraight line ▬▬▬▬, make ▬▬▬ = ▬▬▬, draw ▬▬▬ ⊥ ▬▬▬ ▬·▬·▬ ; biſect ▬▬▬▬, and the point of biſection is the centre.

For, if it be poſſible, let any other point as the point of concourſe of ▬▬▬ , ▬·▬·▬ and ▬·▬·▬·▬ be the centre.

Becauſe in ◹ and ◸

▬▬▬ = ▬·▬·▬ (hyp. and B. 1, def. 15.)

▬▬▬ = ▬·▬·▬ (conſt.) and ▬·▬·▬ common,

◣ = ◢ (B. 1, pr. 8.), and are therefore right

angles ; but ◢ = ◟ (conſt.) ◢ = ◢ (ax. 11.)

which is abſurd ; therefore the aſſumed point is not the centre of the circle ; and in the ſame manner it can be proved that no other point which is not on ▬▬▬ is the centre, therefore the centre is in ▬▬▬, and therefore the point where ▬▬▬ is biſected is the centre.

Q. E. D.

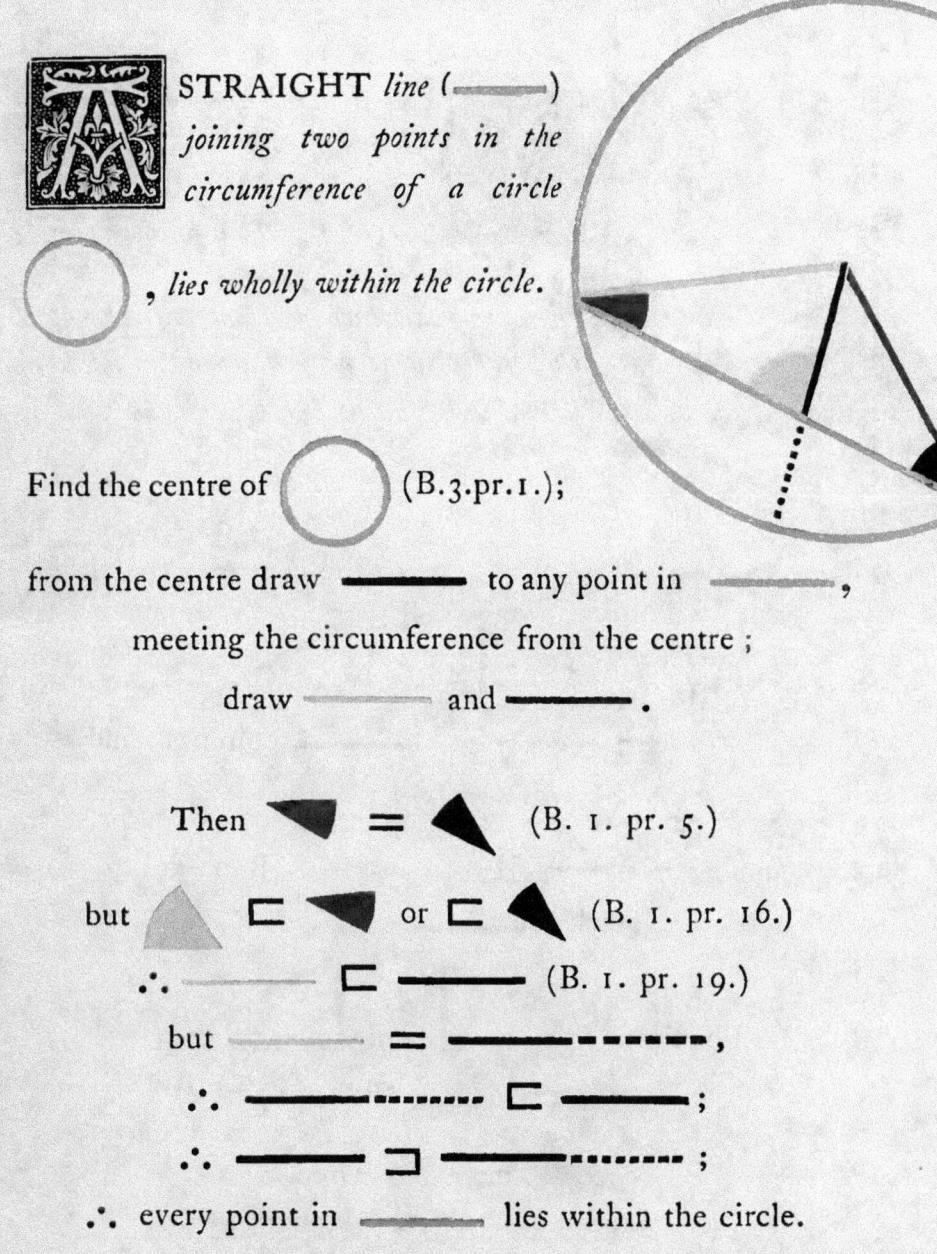

STRAIGHT *line* (———) *joining two points in the circumference of a circle* ⬤ *, lies wholly within the circle.*

Find the centre of ⬤ (B.3.pr.1.);

from the centre draw ▬▬ to any point in ▬▬ ,
meeting the circumference from the centre ;
draw ▬▬ and ▬▬ .

Then ◣ = ◢ (B. 1. pr. 5.)

but ◣ ⊏ ◢ or ⊏ ◢ (B. 1. pr. 16.)

∴ ——— ⊏ —— (B. 1. pr. 19.)

but ——— = ▬▬ ┅ ▬ ,

∴ ▬ ┉ ┅ ⊏ ▬ ;

∴ ▬ ▬ ⊐ ▬ ┅ ;

∴ every point in ——— lies within the circle.

Q. E. D.

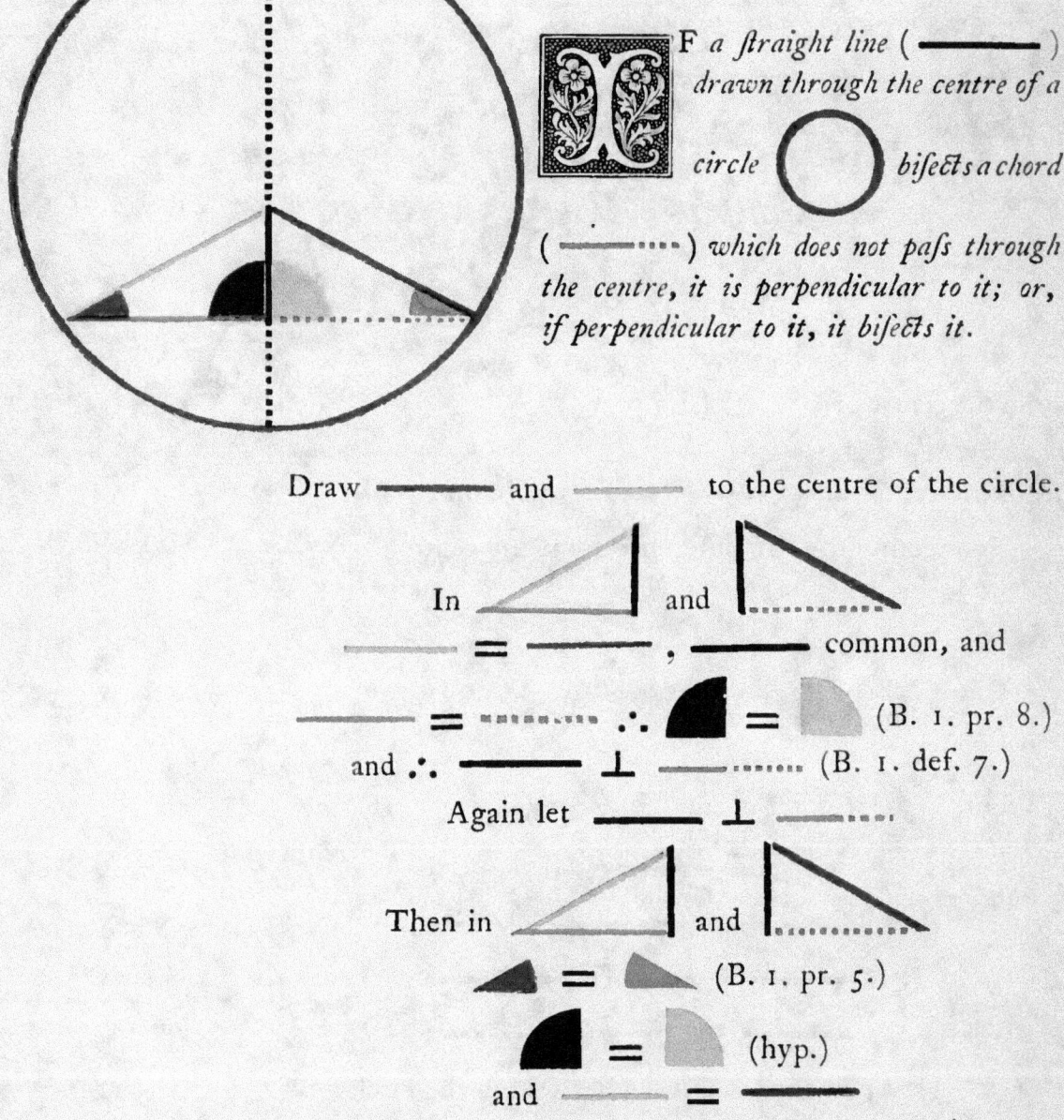

I F a straight line (————) drawn through the centre of a circle ◯ biſects a chord (—— ····) which does not paſs through the centre, it is perpendicular to it; or, if perpendicular to it, it biſects it.

Draw ———— and ———— to the centre of the circle.

In △ and △, ———— = ————, ———— common, and ———— = ··········· ∴ ◢ = ◢ (B. 1. pr. 8.) and ∴ ———— ⊥ ——···· (B. 1. def. 7.)

Again let ———— ⊥ ——····

Then in △ and △ ◣ = ◣ (B. 1. pr. 5.) ◣ = ◣ (hyp.) and ———— = ———— ∴ ———— = ——···· (B. 1. pr. 26.) and ∴ ———— biſects ——···· .

Q. E. D.

 F *in a circle two straight lines cut one another, which do not both pass through the centre, they do not bisect one* another.

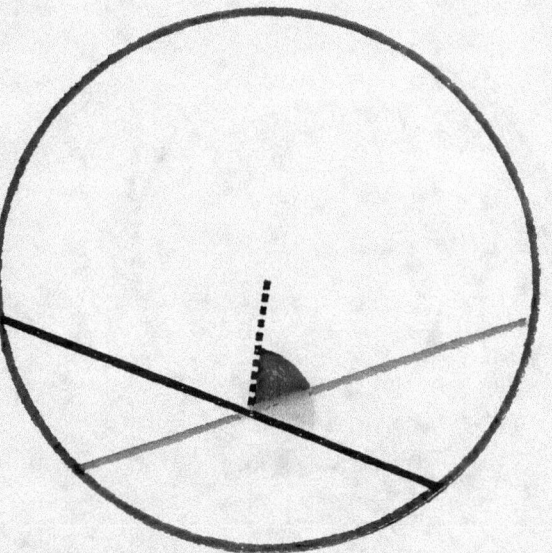

If one of the lines pass through the centre, it is evident that it cannot be bisected by the other, which does not pass through the centre.

But if neither of the lines ——— or ———

pass through the centre, draw ━━━━━━━

from the centre to their intersection.

If ——— be bisected, ━━━━━ ⊥ to it (B. 3. pr. 3.)

∴ ◗ = ◗ and if ——— be

bisected, ━━━━━ ⊥ ——— (B. 3. pr. 3.)

∴ ◗ = ◗

and ∴ ◗ = ◗ ; a part

equal to the whole, which is absurd:

∴ ——— and ———

do not bisect one another.

Q. E. D.

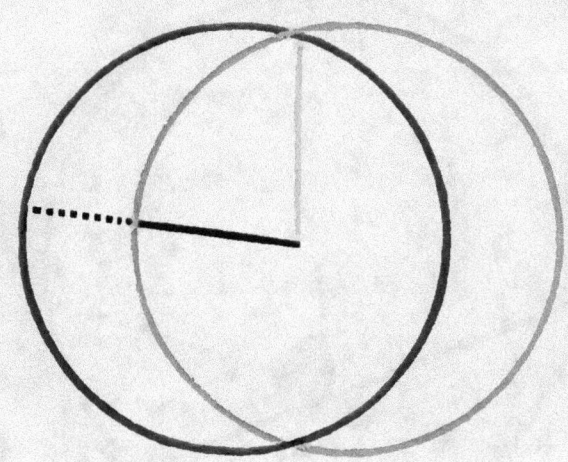

 F *two circles*

interfect, they have not the

fame centre.

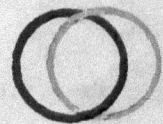

Suppofe it poffible that two interfecting circles have a
common centre; from fuch fuppofed centre draw ————
to the interfecting point, and ————————········· meeting
the circumferences of the circles.

Then ———— = ———— (B. 1. def. 15.)

and ———— = ————————····· (B. 1. def. 15.)

∴ ——— —— = ————————·······; a part

equal to the whole, which is abfurd:

∴ circles fuppofed to interfect in any point cannot

have the fame centre.

Q. E. D.

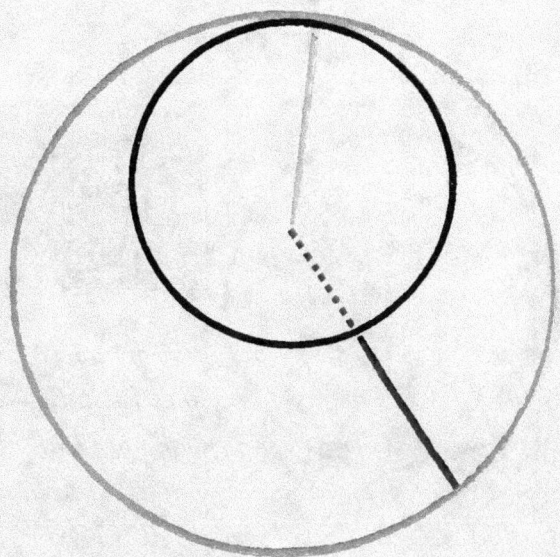

F *two circles*  *touch* *one another internally, they have not the same centre.*

For, if it be poffible, let both circles have the fame centre; from such a fuppofed centre draw ▬▬▬▬ cutting both circles, and ▬▬▬ to the point of contact.

Then ▬▬▬ = ▬▬▬▬ (B. 1. def. 15.)

and ▬▬▬ = ▬▬▬▬ (B. 1. def. 15.)

∴ ▬▬▬▬ = ▬▬▬▬; a part equal to the whole, which is abfurd;

therefore the affumed point is not the centre of both circles; and in the fame manner it can be demonftrated that no other point is.

Q. E. D.

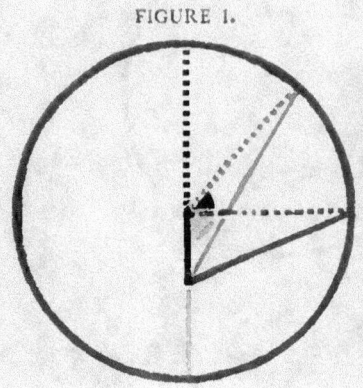

FIGURE I.

I F *from any point within a circle which is not the centre, lines* {

are drawn to the circumference; the greatest of those lines is that (——••••••• *) which passes through the centre, and the least is the remaining part (* —— *) of the diameter.*

Of the others, that (—— *) which is nearer to the line passing through the centre, is greater than that (* —— *) which is more remote.*

Fig. 2. The two lines (——•••• *and* —— *) which make equal angles with that passing through the centre, on opposite sides of it, are equal to each other; and there cannot be drawn a third line equal to them, from the same point to the circumference.*

FIGURE II.

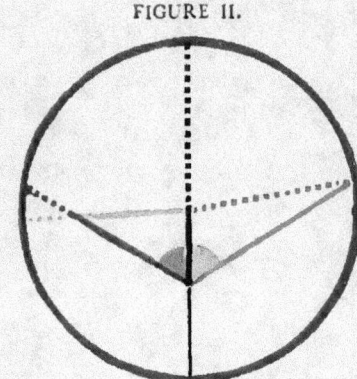

FIGURE I.

To the centre of the circle draw ------- and -------;
then •••••••• = •••••••• (B. 1. def. 15.)
•••——— = —— + ••••••• ⊏ —— (B. 1. pr. 20.)
in like manner ••••——— may be shewn to be greater than
——, or any other line drawn from the same point
to the circumference. Again, by (B. 1. pr. 20.)
—— + —— ⊏ ------- = —— + ——,
take —— from both; ∴ ——— ⊏ —— (ax.),
and in like manner it may be shewn that —— is less

than any other line drawn from the same point to the cir-

cumference. Again, in 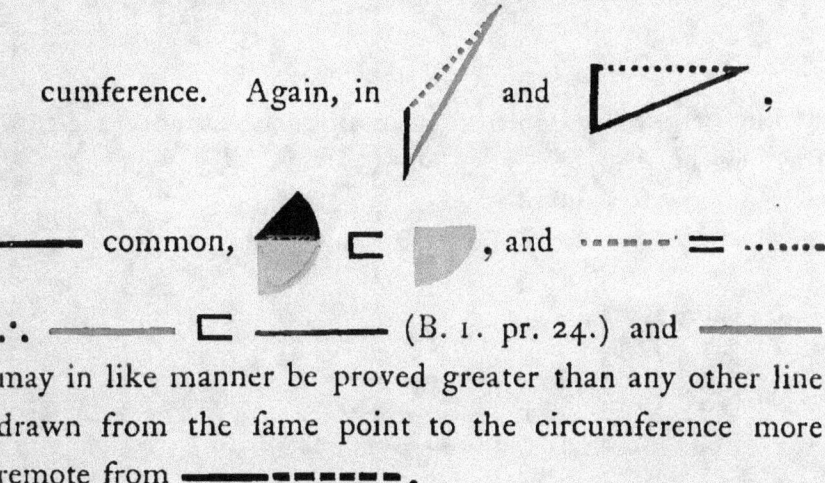 and ;

—— common, ⊏ , and ⋯⋯ ≡ ⋯⋯

∴ —— ⊏ ———— (B. 1. pr. 24.) and ————

may in like manner be proved greater than any other line
drawn from the same point to the circumference more
remote from ■■■■■.

FIGURE II.

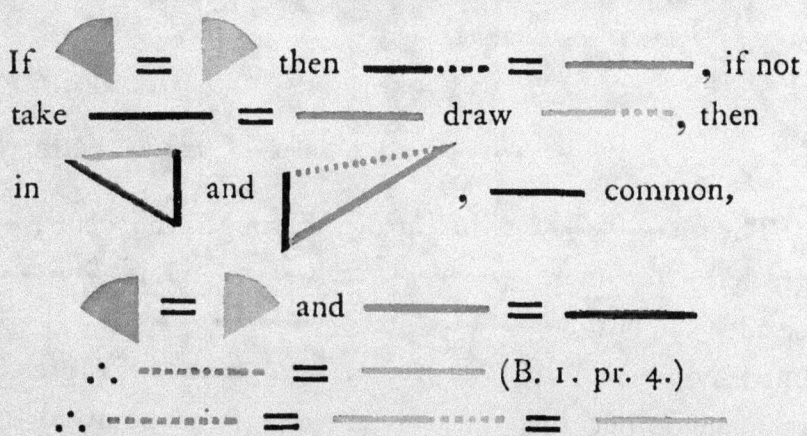

If ◀ = ▶ then ——⋯ = ———, if not

take ——— = ——— draw ⋯⋯, then

in and , —— common,

◀ = ▶ and ——— = ———

∴ ⋯⋯ = ——— (B. 1. pr. 4.)

∴ ⋯⋯ = ——— = ———

a part equal to the whole, which is abfurd:

∴ ——— = ———⋯; and no other line is equal to
——— drawn from the same point to the circumference; for if it were nearer to the one pafing through the
centre it would be greater, and if it were more remote it
would be lefs.

Q. E. D.

M

The original text of this propofition is here divided into three parts.

I.

IF *from a point without a circle, ftraight* lines $\left\{ \begin{matrix} \rule{1cm}{0.4pt}\cdots \\ \rule{1cm}{1pt} \\ \rule{1cm}{1pt}\ \&c. \end{matrix} \right\}$ *are drawn to the cir-* cumference; *of thofe falling upon the concave circum-* ference the greateft is that (⸻••••) *which paffes* through the centre, and the line (⸻) *which is* nearer the greateft is greater than that (⸻) which is more remote.

Draw ••••••••• and ••••••••• to the centre.

Then, ⸻•••• which paffes through the centre, is greateft; for fince •••••••• = •••••••• , if ⸻ be added to both, ⸻•••• = ⸻ + ••••••••• ; but ⊏ ⸻ (B. 1. pr. 20.) ∴ ⸻•••• is greater than any other line drawn from the fame point to the concave circumference.

Again in 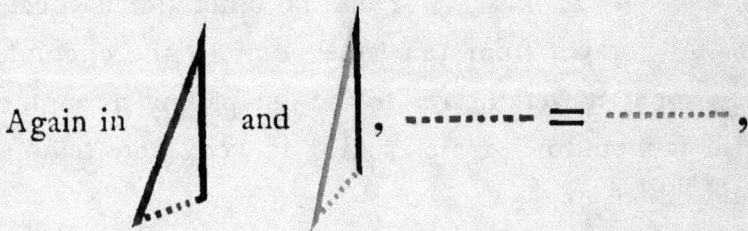 and , •••••••• = ••••••••• ,

and —————— common, but ▰ ⊏ ▱ ,

∴ ————— ⊏ —————— (B. 1. pr. 24.) ;

and in like manner ————— may be ſhewn ⊏ than any

other line more remote from ——————.

II.

Of thoſe lines falling on the convex circumference the leaſt is that (----------) which being produced would paſs through the centre, and the line which is nearer to the leaſt is leſs than that which is more remote.

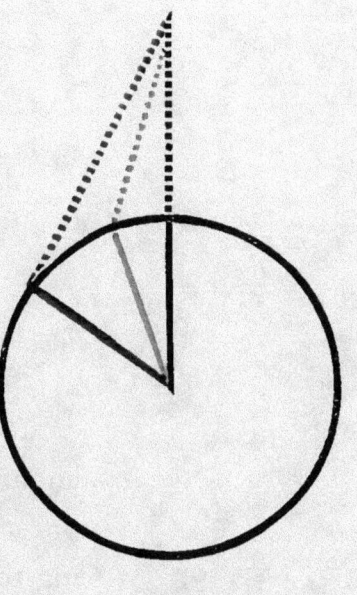

For, ſince ———— + -------- ⊏ ————•••• (B. 1. pr. 20.)

and ————— = ————— ,

∴ -------- ⊏ ————•••••• (ax. 5.)

And again, ſince ————— + ------- ⊏

————— + ------- (B. 1. pr. 21.),

and ————— = ————— ,

∴ -------- ⊐ -------•. And ſo of others.

III.

Alſo the lines making equal angles with that which paſſes through the centre are equal, whether falling on the concave or convex circumference ; and no third line can be drawn equal to them from the ſame point to the circumference.

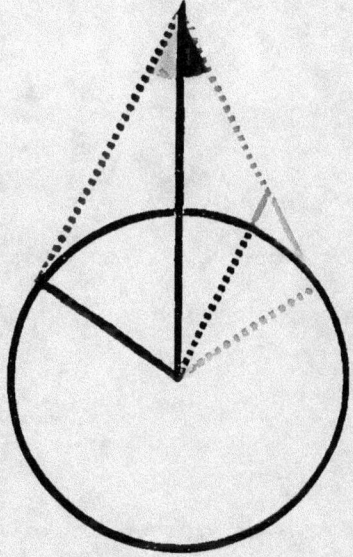

For if -------- ⊏ -------, but making ◣ = ◤ ;

make -------- = -------, and draw -------.

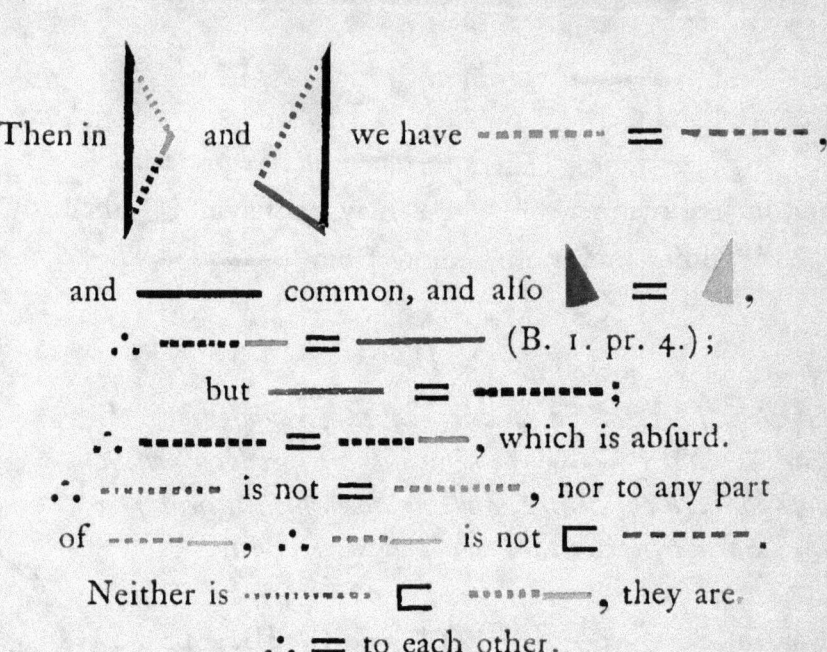

Then in ▷ and ◁ we have ▬▬▬ $=$ ▬▬▬ ,

and ▬▬▬ common, and also ◣ $=$ ◤ ,

∴ ▬▬▬ $=$ ▬▬▬ (B. 1. pr. 4.);

but ▬▬▬ $=$ ▬▬▬ ;

∴ ▬▬▬ $=$ ▬▬▬ , which is abſurd.

∴ ▬▬▬ is not $=$ ▬▬▬ , nor to any part

of ▬▬▬ , ∴ ▬▬▬ is not $\sqsubset$ ▬▬▬ .

Neither is ▬▬▬ $\sqsubset$ ▬▬▬ , they are

∴ $=$ to each other.

And any other line drawn from the ſame point to the
circumference muſt lie at the ſame ſide with one of theſe
lines, and be more or leſs remote than it from the line paſſing through the centre, and cannot therefore be equal to it.

Q. E. D.

F *a point be taken within a circle* ◯ , *from which more than two equal straight lines* (▬▬ , ▬▬ , ▬▬) *can be drawn to the circumference, that point muſt be the centre of the circle.*

For, if it be ſuppoſed that the point ╲ in which more than two equal ſtraight lines meet is not the centre, ſome other point ▬ muſt be; join theſe two points by ▬▬ , and produce it both ways to the circumference.

Then ſince more than two equal ſtraight lines are drawn from a point which is not the centre, to the circumference, two of them at leaſt muſt lie at the ſame ſide of the diameter ▬▬▬ ; and ſince from a point ◿◺ , which is not the centre, ſtraight lines are drawn to the circumference; the greateſt is ▬▬ , which paſſes through the centre : and ▬▬ which is nearer to ▬▬ , ⊏▬ which is more remote (B. 3. pr. 8.); but ▬▬ ≡ ▬▬ (hyp.) which is abſurd.

The ſame may be demonſtrated of any other point, different from ◿◺ , which muſt be the centre of the circle.

Q. E. D.

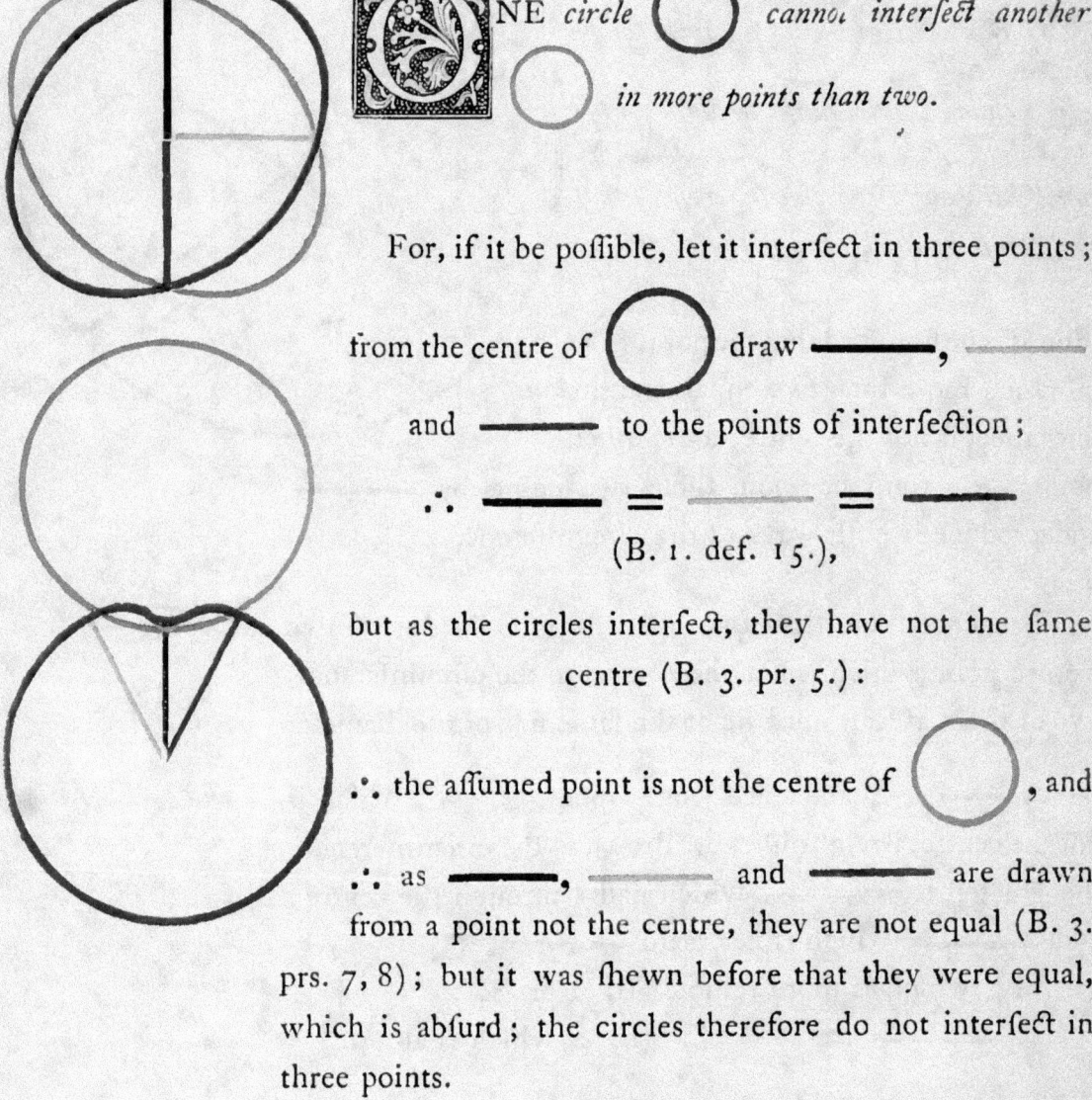

ONE *circle* ⬭ *cannot intersect another in more points than two.*

For, if it be possible, let it intersect in three points;

from the centre of ⬭ draw ▬▬▬, ▬▬▬

and ▬▬▬ to the points of intersection;

∴ ▬▬▬ = ▬▬▬ = ▬▬▬

(B. 1. def. 15.),

but as the circles intersect, they have not the same centre (B. 3. pr. 5.):

∴ the assumed point is not the centre of ⬭, and

∴ as ▬▬▬, ▬▬▬ and ▬▬▬ are drawn from a point not the centre, they are not equal (B. 3. prs. 7, 8); but it was shewn before that they were equal, which is absurd; the circles therefore do not intersect in three points.

Q. E. D.

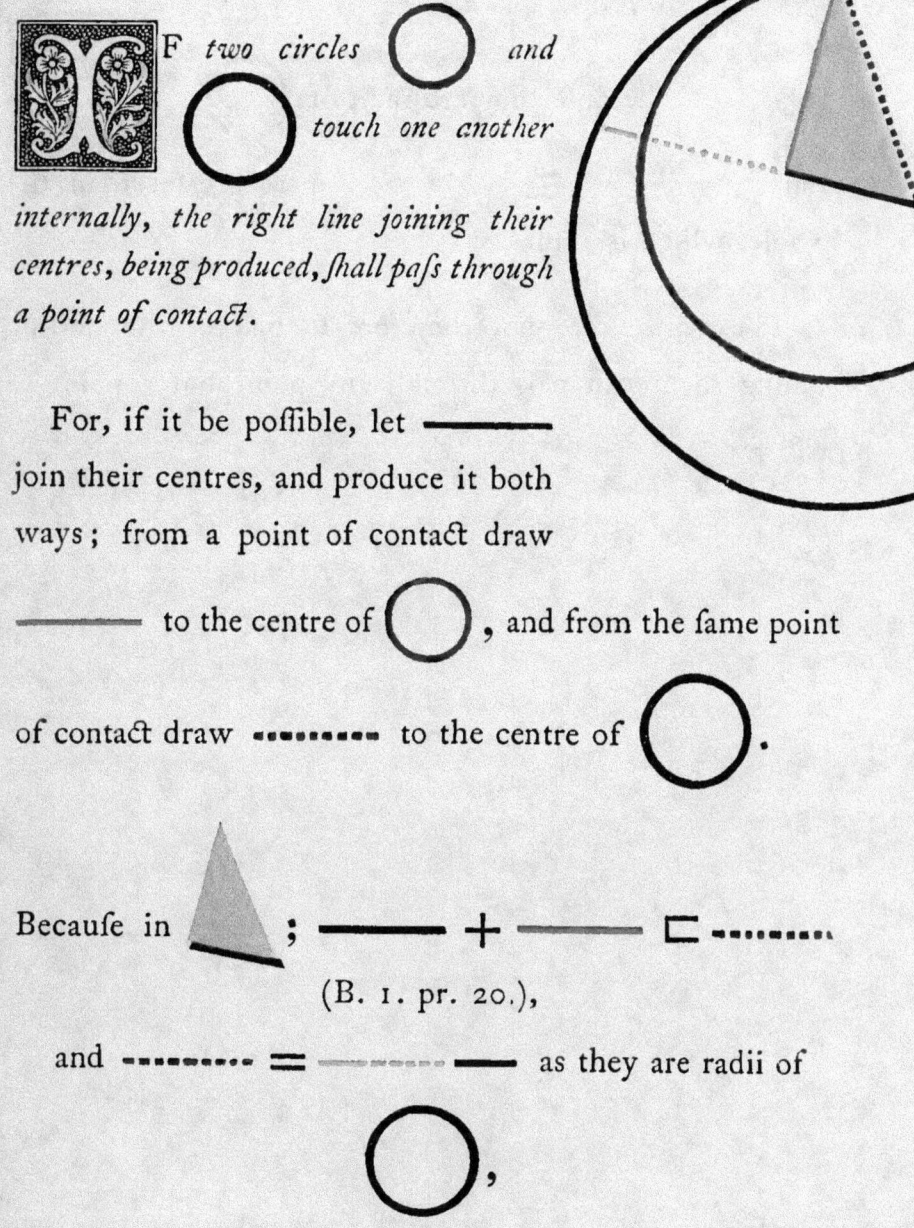

F *two circles* 〇 *and* 〇 *touch one another internally, the right line joining their centres, being produced, shall pass through a point of contact.*

For, if it be possible, let —— join their centres, and produce it both ways; from a point of contact draw

—— to the centre of 〇, and from the same point

of contact draw ••••••••• to the centre of 〇.

Because in ◢ ; —— + —— ⊏ •••••••
(B. 1. pr. 20.),

and ••••••••• = —————— as they are radii of

〇,

but ▬ **+** ▬▬▬ ⊏ ▬▬▬▬ ; take

away ▬▬▬ which is common,

and ▬▬▬ ⊏ ▬▬▬▬▬ ;

but ▬▬▬ **=** ▬▬▬▬ ,

because they are radii of ◯ ,

and ∴ ▬▬▬▬ ⊏ ▬▬▬▬▬ a part greater than the whole, which is abfurd.

The centres are not therefore fo placed, that a line joining them can pafs through any point but a point of contact.

Q. E. D.

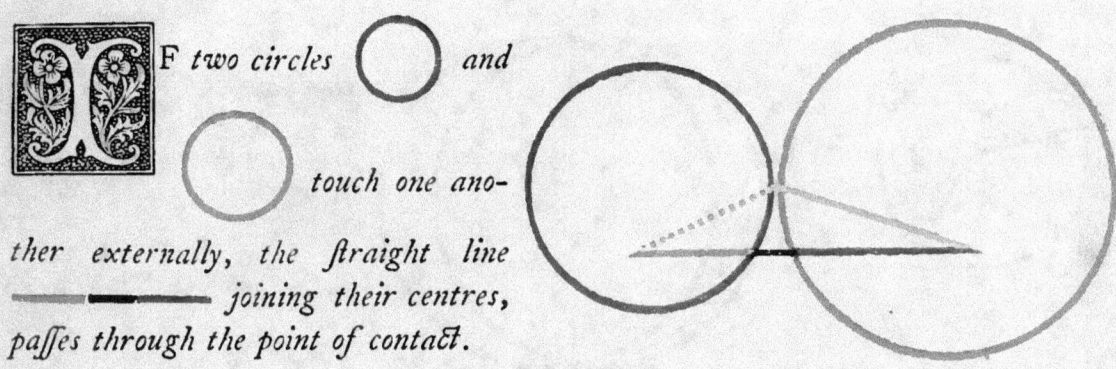

F *two circles* ⬭ *and* ⬭ *touch one another externally, the straight line* ▬▬▬ *joining their centres, passes through the point of contact.*

If it be possible, let ▬▬▬ join the centres, and not pass through a point of contact; then from a point of contact draw ----- and ▬▬ to the centres.

Because ------ $+$ ▬▬ $\sqsubset$ ▬▬▬ (B. 1. pr. 20.),

and ▬▬ $=$ ----- (B. 1. def. 15.),

and ▬▬ $=$ ▬▬ (B. 1. def. 15.),

∴ ▬ $+$ ▬ $\sqsubset$ ▬▬▬, a part greater than the whole, which is abfurd.

The centres are not therefore fo placed, that the line joining them can pafs through any point but the point of contact.

Q. E. D.

N

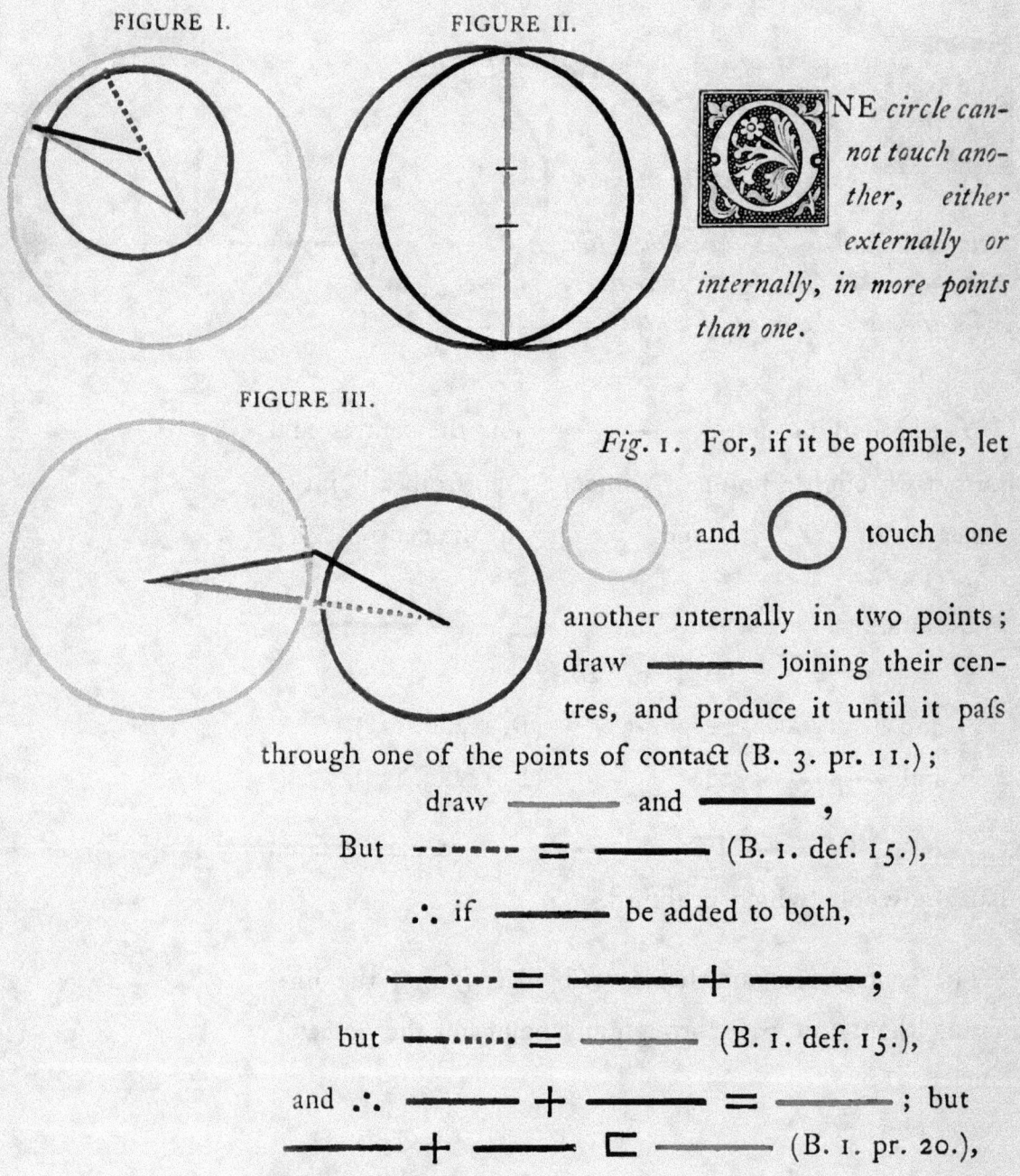

FIGURE I.

FIGURE II.

FIGURE III.

ONE *circle can-not touch ano-ther, either externally or internally, in more points than one.*

Fig. 1. For, if it be poffible, let

and   touch   one

another internally in two points; draw ——— joining their cen-tres, and produce it until it pafs through one of the points of contact (B. 3. pr. 11.);

draw ——— and ———,

But ------- ⚌ ——— (B. 1. def. 15.),

∴ if ——— be added to both,

———···· ⚌ ——— + ———;

but ———···· ⚌ ——— (B. 1. def. 15.),

and ∴ ——— + ——— = ———; but

——— + ——— ⊏ ——— (B. 1. pr. 20.),

which is abfurd.

Fig. 2.   But if the points of contact be the extremities of the right line joining the centres, this ſtraight line muſt be biſected in two different points for the two centres; becauſe it is the diameter of both circles, which is abſurd.

Fig. 3.   Next, if it be poſſible, let 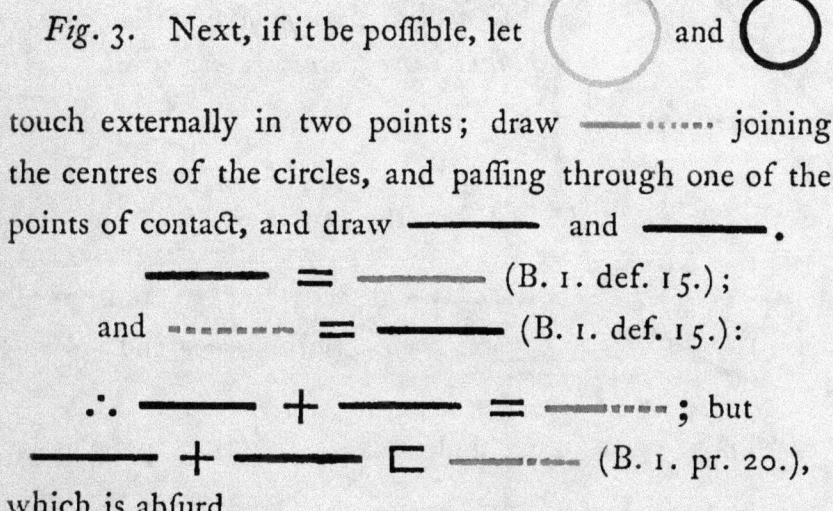 and

touch externally in two points; draw ——······· joining the centres of the circles, and paſſing through one of the points of contact, and draw ——— and ———.

———— = ———— (B. 1. def. 15.);

and ········· = ———— (B. 1. def. 15.):

∴ —— + —— = ········· ; but

—— + —— ⊏ ········· (B. 1. pr. 20.),

which is abſurd.

There is therefore no caſe in which two circles can touch one another in two points.

Q E. D.

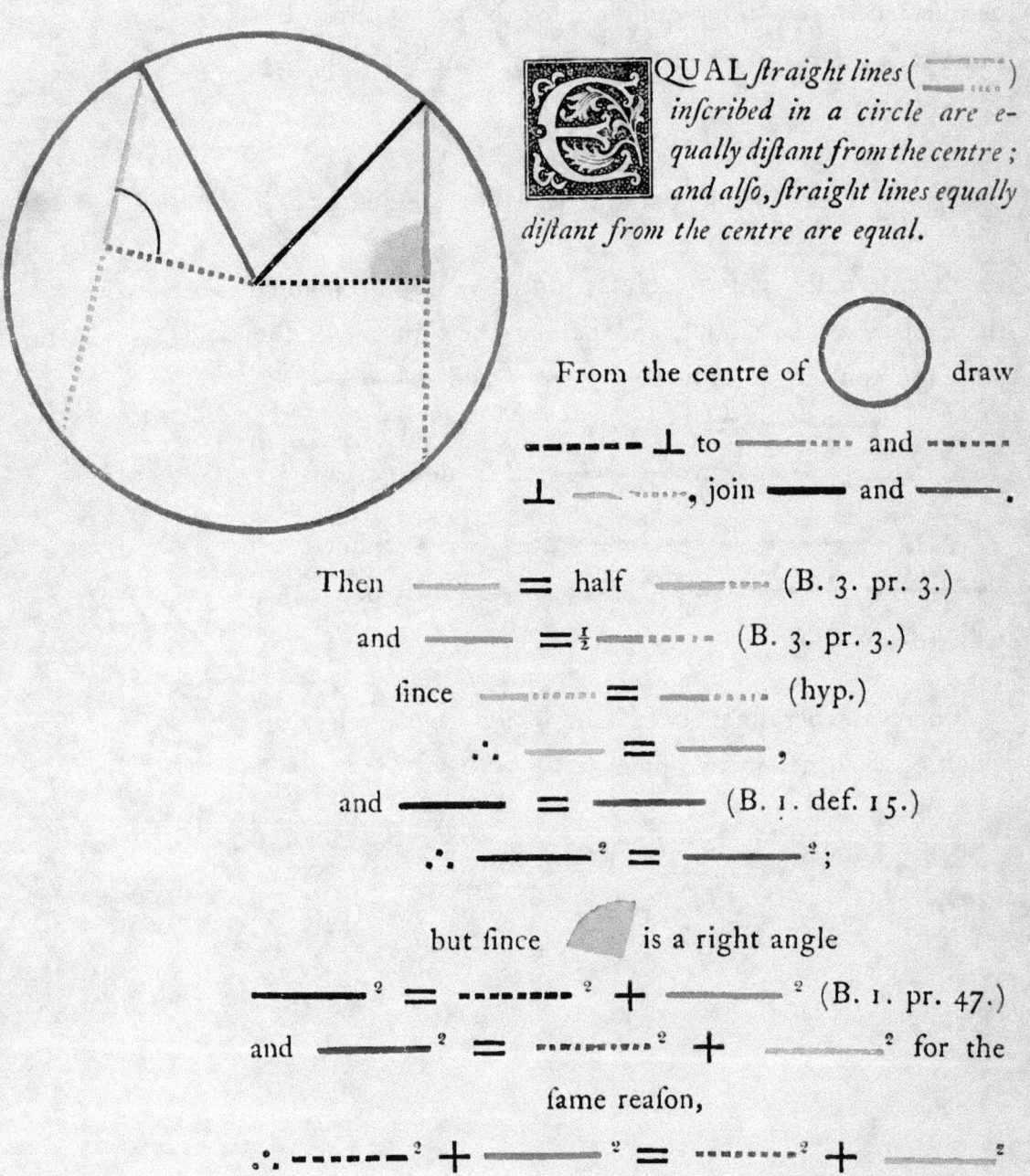

EQUAL *straight lines* (━━━)
inscribed in a circle are e-
qually diſtant from the centre;
and alſo, ſtraight lines equally
diſtant from the centre are equal.

From the centre of ◯ draw

━ ━ ━ ⊥ to ━━━ and ╍╍╍

⊥ ━━━, join ━━━ and ━━━.

Then ━━━ = half ━━━ (B. 3. pr. 3.)

and ━━━ = ½ ━━━ (B. 3. pr. 3.)

ſince ━━━ = ━━━ (hyp.)

∴ ━━━ = ━━━,

and ━━━ = ━━━ (B. 1. def. 15.)

∴ ━━━² = ━━━²;

but ſince ◢ is a right angle

━━━² = ╍╍╍² + ━━━² (B. 1. pr. 47.)

and ━━━² = ╍╍╍² + ━━━² for the

ſame reaſon,

∴ ╍╍╍² + ━━━² = ╍╍╍² + ━━━²

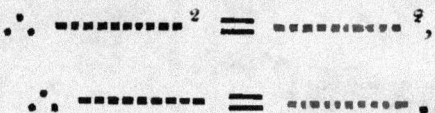

$$\therefore \text{------}^2 = \text{------}^2,$$

$$\therefore \text{------} = \text{------}.$$

Alſo, if the lines ——— and ——— be equally diſtant from the centre; that is to ſay, if the perpendiculars ——— and ——— be given equal, then ——— = ———.

For, as in the preceding caſe,

$$\text{------}^2 + \text{------}^2 = \text{------}^2 + \text{------}^2;$$

but $\text{------}^2 = \text{------}^2$:

$\therefore \text{------}^2 = \text{------}^2$, and the doubles of theſe ——— and ——— are alſo equal.

Q. E. D.

FIGURE I.

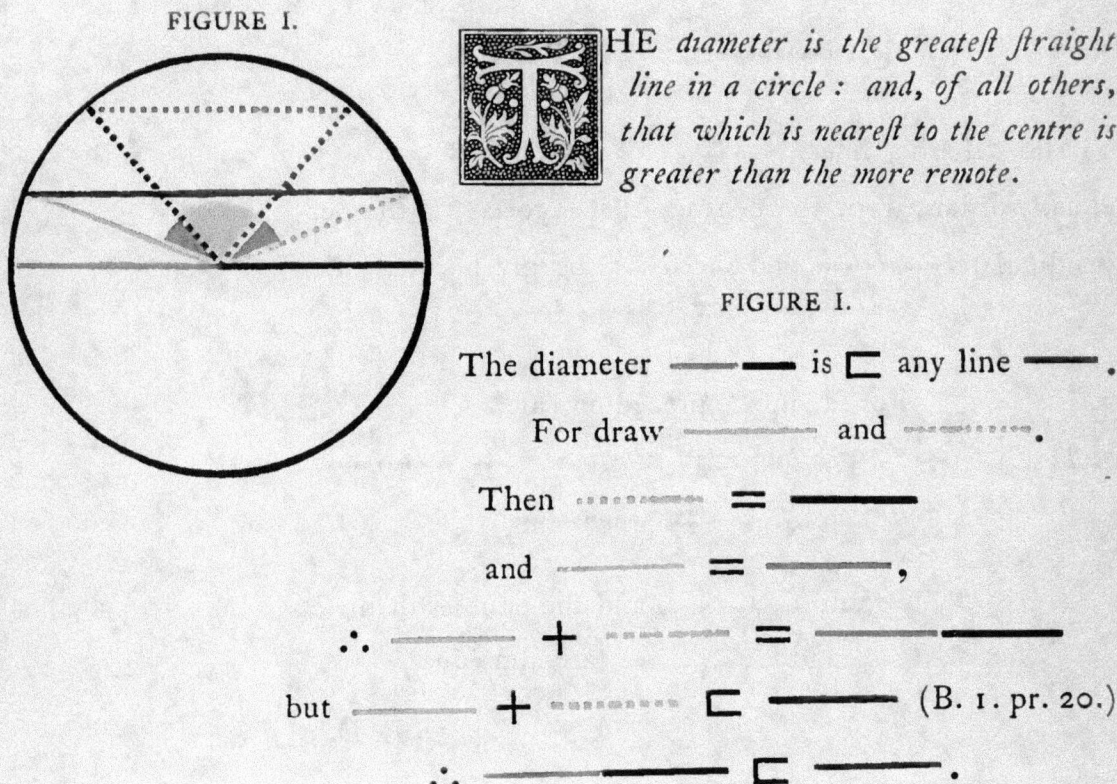

THE *diameter is the greateſt ſtraight line in a circle : and, of all others, that which is neareſt to the centre is greater than the more remote.*

FIGURE I.

The diameter ——— ━━ is ⊏ any line ——— .

For draw ——— and ·········· .

Then ·········· = ━━━━

and ——— = ——— ,

∴ ——— ✛ ·········· = ——— ━━━━

but ——— ✛ ·········· ⊏ ━━━   (B. 1. pr. 20.)

∴ ——— ━━ ⊏ ━━━ .

Again, the line which is nearer the centre is greater than the one more remote.

Firſt, let the given lines be ━━━ and --------- , which are at the ſame ſide of the centre and do not interſect ;

draw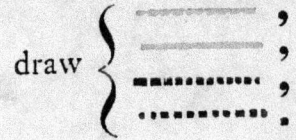

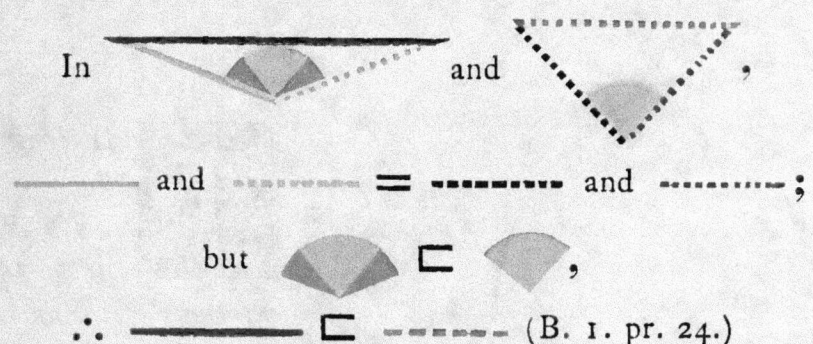

In   and   ,

———— and ·······－ ≡ ▬▬▬▬ and ·········· ;

but ◢◣ ⊏ ◥ ,

∴ ▬▬▬▬ ⊏ ·－·－·   (B. 1. pr. 24.)

FIGURE II.

Let the given lines be ———— and ———

which either are at different fides of the centre,

or interfect ; from the centre draw ·－·－·－·

and ▬ ▬ ▬ ⊥ ———— and ———— ,

make ········ ≡ ·－·－ , and

draw ▬▬▬▬ ⊥ ·－·－· .

Since ———— and ▬▬▬▬ are equally diftant from

the centre, ———— ≡ ▬▬▬▬ (B. 3. pr. 14.);

but ▬▬▬▬ ⊏ ———— (Pt. 1. B. 3. pr. 15.),

∴ ———— ⊏ ———— .

Q. E. D.

FIGURE II.

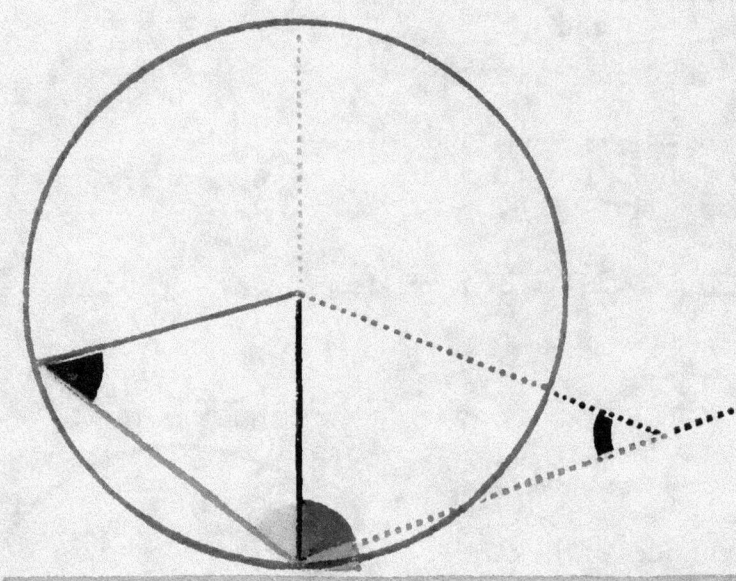

HE *ſtraight line* ——— *drawn from the extremity of the diameter* —— *of a circle perpendicular to it falls without the circle.*

And if any ſtraight line - - - - - *be drawn from a point within that perpendicular to the point of contact, it cuts the circle.*

PART I

If it be poſſible, let ———, which meets the circle again, be ⊥ ———, and draw ———.

Then, becauſe ——— = ———,

◣ = ◢ (B. 1. pr. 5.),

and ∴ each of these angles is acute. (B. 1. pr. 17.)

but ◣ = ◗ (hyp.), which is abſurd, therefore ——— drawn ⊥ ——— does not meet the circle again.

PART II.

Let ▬▬▬ be ⊥ ▬▬ and let ▬▬▬ be drawn from a point ⟋ between ▬▬▬ and the circle, which, if it be poſſible, does not cut the circle.

Becauſe ◗ = ◗ ,

∴ ◗ is an acute angle ; ſuppoſe ▬▬▬▬ ⊥ ▬▬▬, drawn from the centre of the circle, it muſt fall at the ſide of ◗ the acute angle.

∴ ◖ which is ſuppoſed to be a right angle, is ⊏ ◗ ,

∴ ▬▬ ⊏ ▬▬▬ ;

but ▬▬▬ = ▬▬ ,

and ∴ ▬▬▬ ⊏ ▬▬▬, a part greater than the whole, which is abſurd. Therefore the point does not fall outſide the circle, and therefore the ſtraight line ▬▬▬ cuts the circle.

Q. E. D.

o

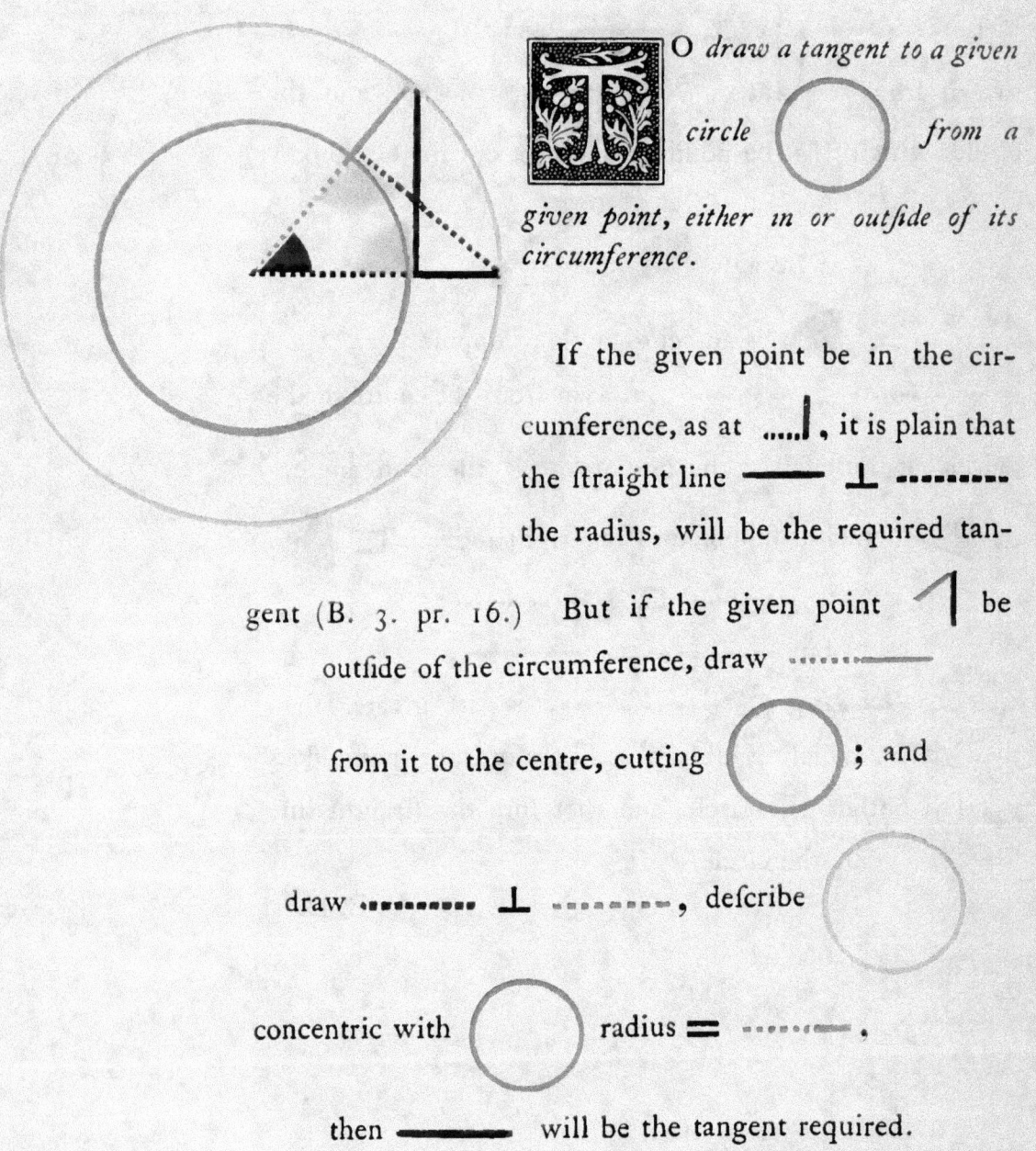

O *draw a tangent to a given circle* 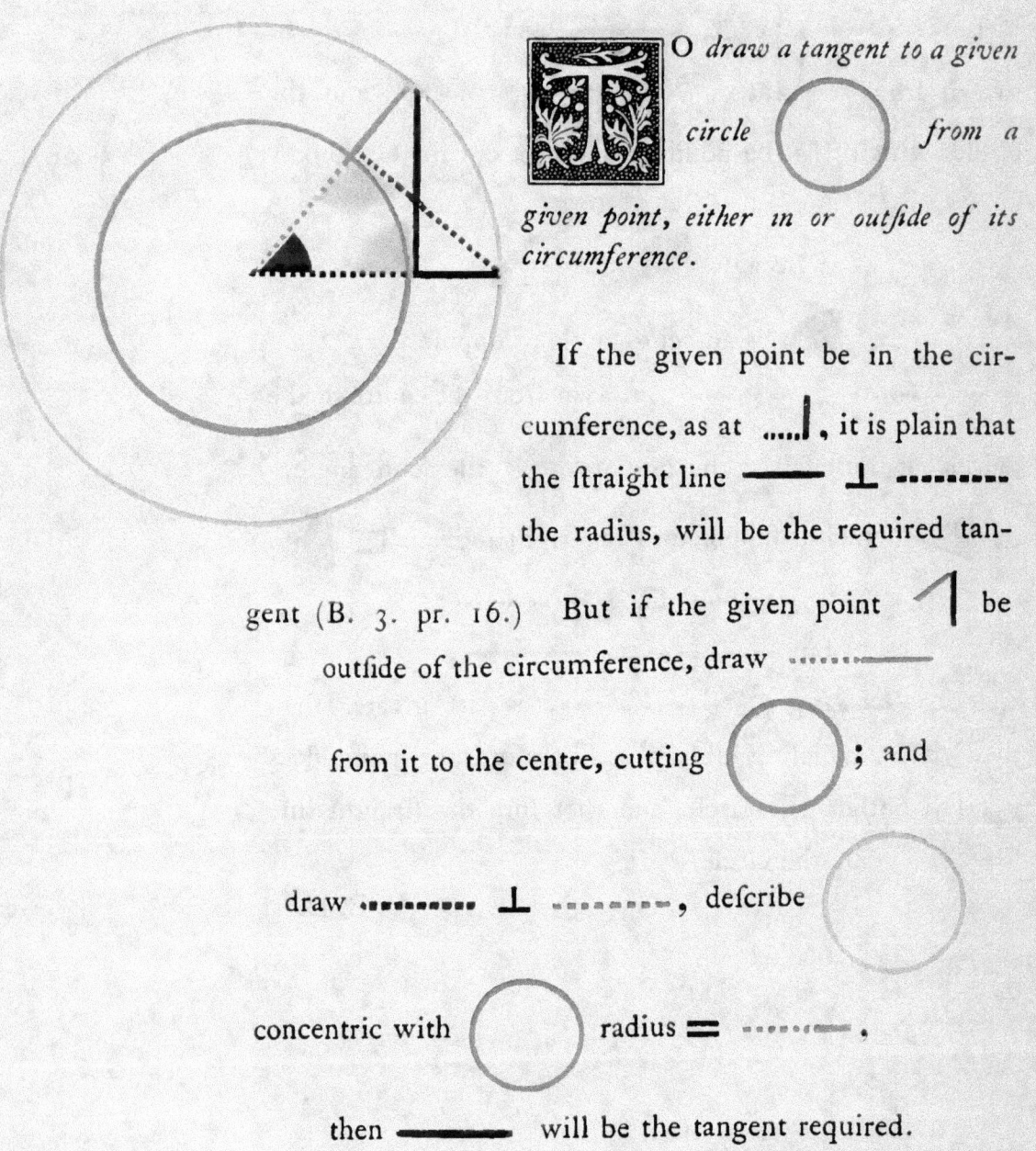 *from a given point, either in or outſide of its circumference.*

If the given point be in the cir-cumference, as at ....⌐ , it is plain that the ſtraight line ▬▬ ⊥ ---------- the radius, will be the required tan-gent (B. 3. pr. 16.) But if the given point ◥ be outſide of the circumference, draw ········▬ from it to the centre, cutting ◯ ; and

draw ·····▬ ⊥ --------- , deſcribe ◯ concentric with ◯ radius ▬ ----- .

then ▬▬ will be the tangent required.

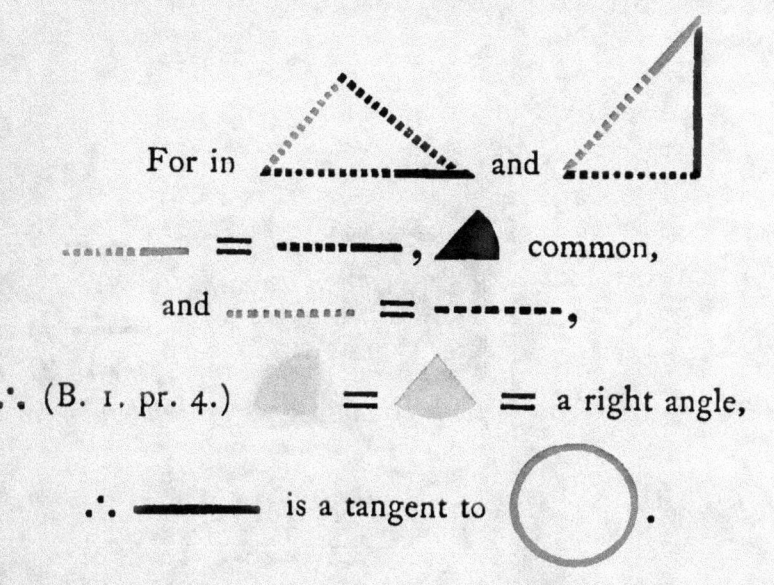

For in ⟨triangle⟩ and ⟨triangle⟩

▬ = ▬, ◢ common,

and ▬ = ▬,

∴ (B. 1. pr. 4.) = = a right angle,

∴ ▬ is a tangent to ◯.

Q. E. D.

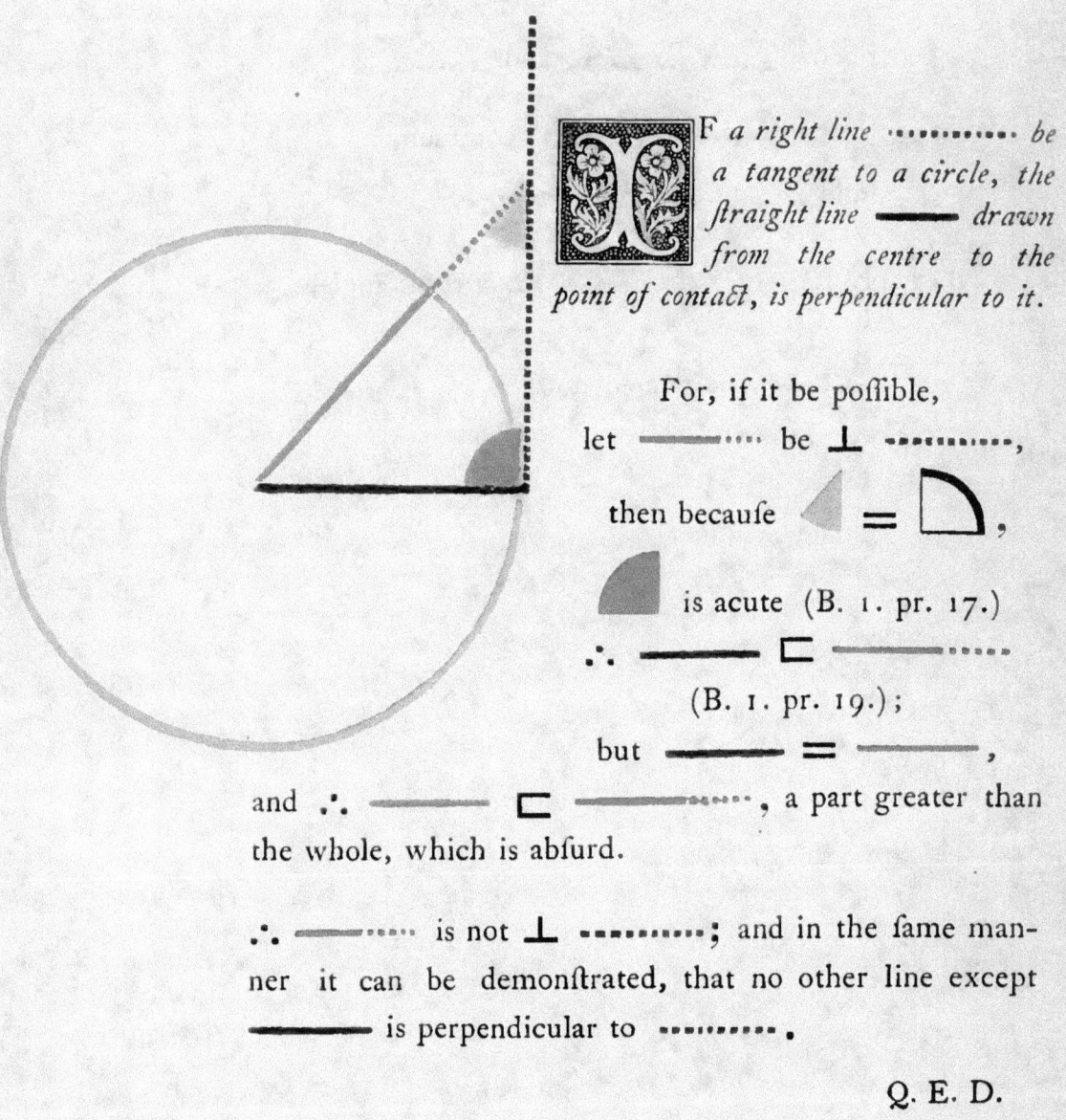

F *a right line* ·········· *be a tangent to a circle, the straight line* —— *drawn from the centre to the point of contact, is perpendicular to it.*

For, if it be poſſible,

let —— ···· be ⊥ ············,

then becauſe ◗ = ◖,

◗ is acute (B. 1. pr. 17.)

∴ —— ⊏ —— ·····

(B. 1. pr. 19.);

but —— = ——,

and ∴ —— ⊏ —— ····, a part greater than the whole, which is abſurd.

∴ —— ···· is not ⊥ ············; and in the ſame manner it can be demonſtrated, that no other line except —— is perpendicular to ··········.

Q. E. D.

IF a ſtraight line be a tangent to a circle, the ſtraight line , drawn perpendicular to it from point of the contact, paſſes through the centre of the circle.

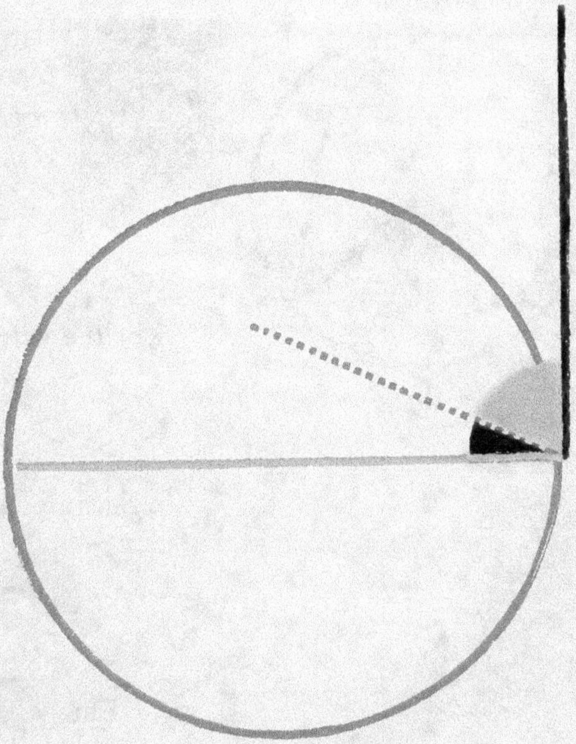

For, if it be poſſible, let the centre be without , and draw from the ſuppoſed centre to the point of contact.

Becauſe ⊥
(B. 3. pr. 18.)

∴ = , a right angle;

but = (hyp.), and ∴ = ,
a part equal to the whole, which is abſurd.

Therefore the aſſumed point is not the centre; and in the ſame manner it can be demonſtrated, that no other point without is the centre.

Q. E. D.

FIGURE I

HE *angle at the centre of a circle, is double the angle at the circumference, when they have the same part of the circumference for their base.*

FIGURE I.

Let the centre of the circle be on ———— ┉┉

a fide of ▲ .

Becaufe ━━━ = ———— ,

▲ = ◣ (B. 1. pr. 5.).

But ◗ = ▲ + ◣ ,

or ◗ = twice ▲ (B. 1. pr. 32).

FIGURE II.

FIGURE II.

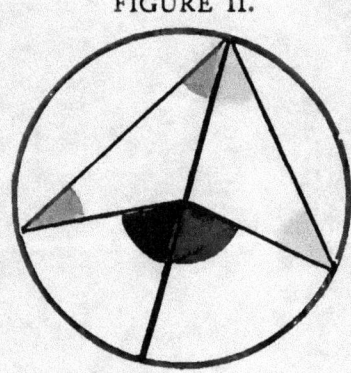

Let the centre be within ◣ , the angle at the circumference ; draw ━━━ from the angular point through the centre of the circle ;

then ◀ = ▶ , and ▲ = ◣ ,

becaufe of the equality of the fides (B. 1. pr. 5).

Hence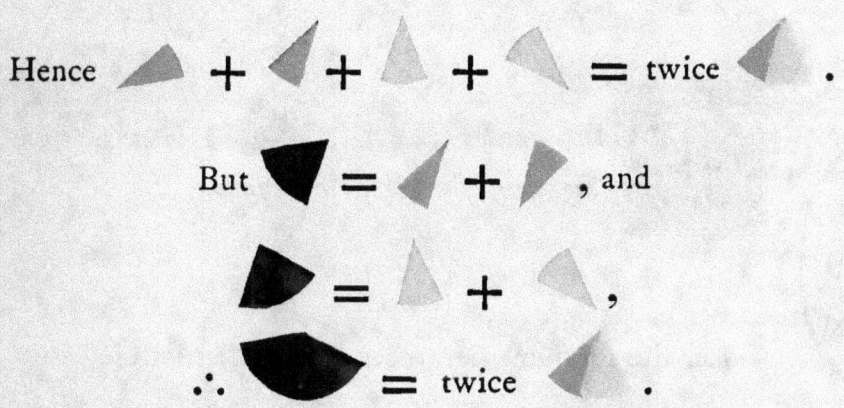

But ▼ = ◄ + ► , and

◗ = ◭ + ◣ ,

∴ ⬤ = twice ◤ .

FIGURE III.

Let the centre be without ◥ and

draw ——————, the diameter.

Because ◢ = twice ◥ ; and

▼ = twice ▼ (cafe 1.) ;

∴ ◭ = twice ◥ .

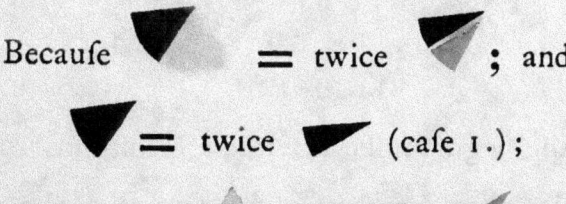

FIGURE III.

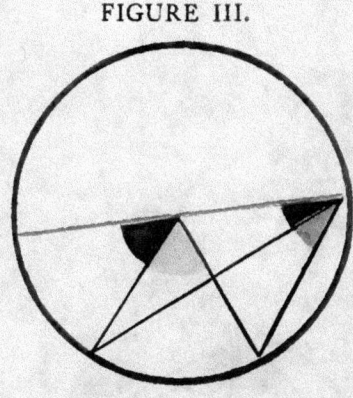

Q. E. D.

FIGURE I.

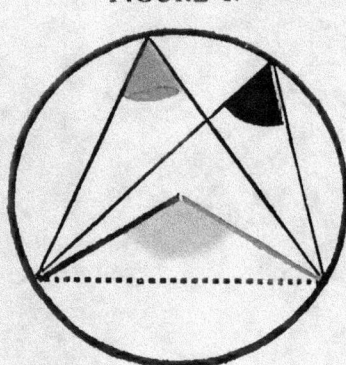

 HE *angles* (,) *in the same segment of a circle are equal.*

FIGURE I.

Let the segment be greater than a semicircle, and draw ———— and ———— to the centre.

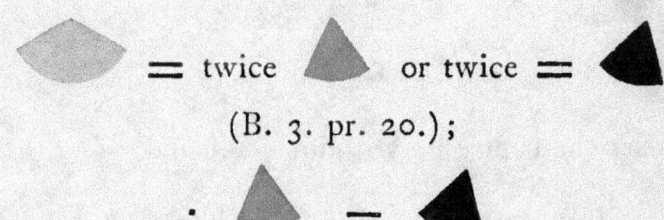

(B. 3. pr. 20.);

$$\therefore \quad \blacktriangle = \blacktriangle .$$

FIGURE II.

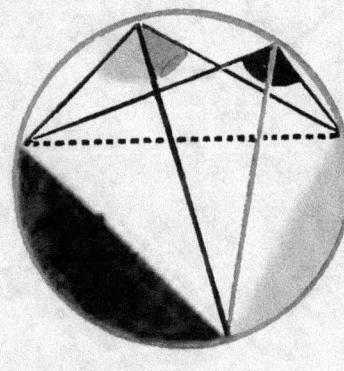

FIGURE II.

Let the segment be a semicircle, or less than a semicircle, draw ———— the diameter, also draw ————.

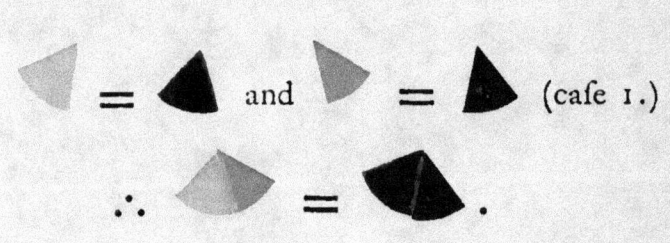

$$\therefore \quad \blacktriangle = \blacktriangle .$$

Q. E. D.

 HE *oppofite angles*

and  , *and*

of any quadrilateral figure in-
fcribed in a circle, are together equal to
two right angles.

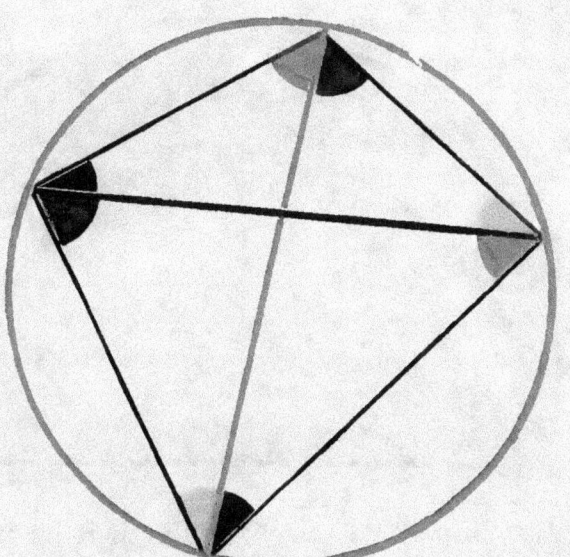

Draw ———— and ———— the diagonals; and becaufe angles in

the fame fegment are equal 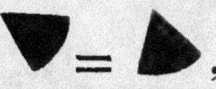 ,

and 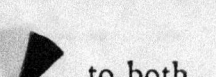 ;

add to both.

∴ two right angles (B. 1. pr. 32.). In like manner it may be fhown that,

Q. E. D.

105

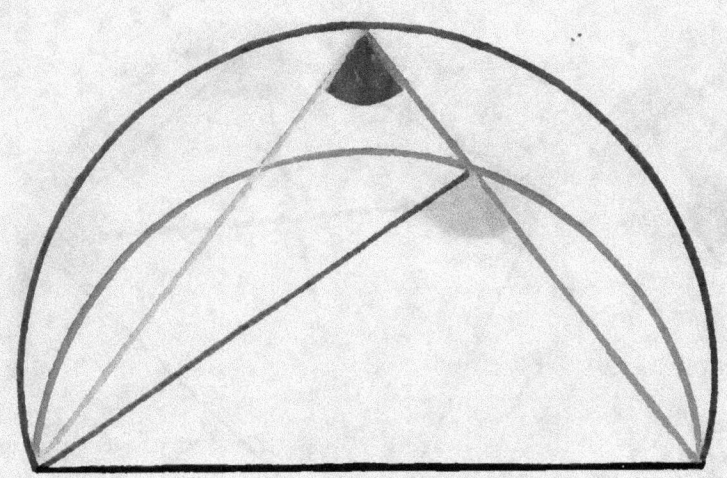

UPON *the same straight line, and upon the same side of it, two similar segments of circles cannot be constructed which do not coincide.*

For if it be poffible, let two fimilar fegments

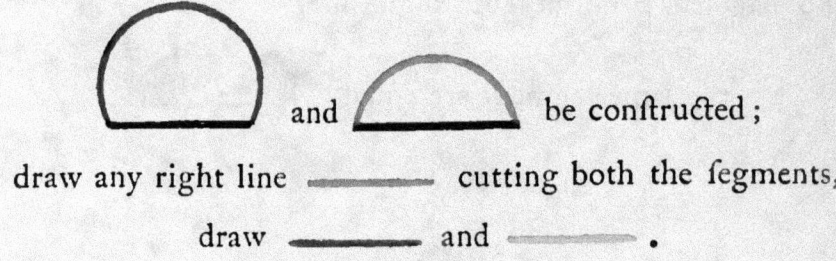

and   be conftructed;

draw any right line ———— cutting both the fegments,

draw ———— and ———— .

Becaufe the fegments are fimilar,

=   (B. 3. def. 10.),

but   ⊏   (B. 1. pr. 16.)

which is abfurd : therefore no point in either of the fegments falls without the other, and therefore the fegments coincide.

Q. E. D.

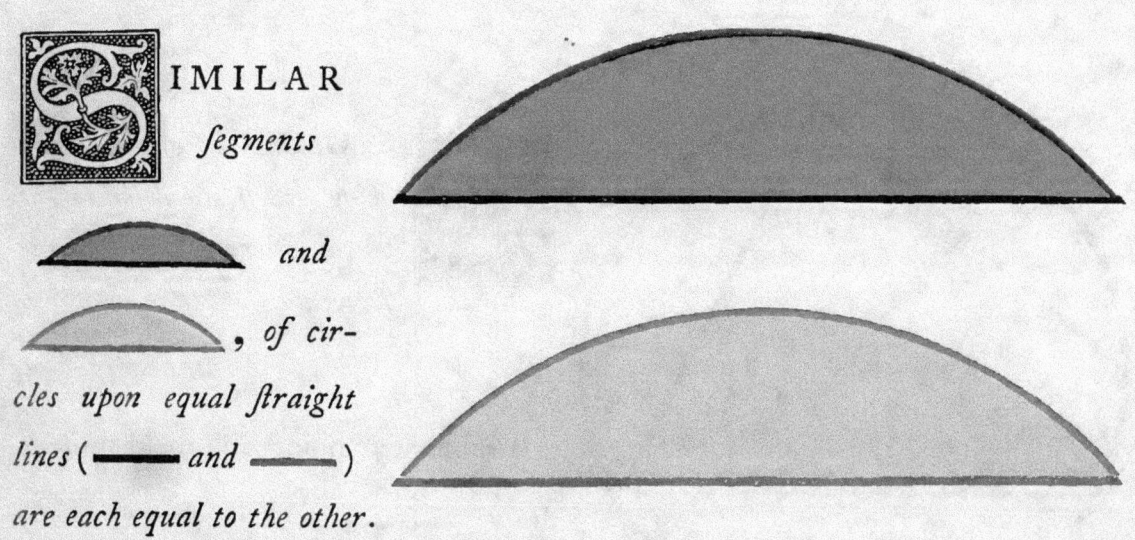

IMILAR *fegments* and , *of circles upon equal ftraight lines* (—— *and* ——) *are each equal to the other.*

For, if be fo applied to , that —— may fall on ——, the extremities of —— may be on the extremities —— and at the fame fide as ;

becaufe —— = ——, —— muft wholly coincide with ——; and the fimilar fegments being then upon the fame ftraight line and at the fame fide of it, muft alfo coincide (B. 3. pr. 23.), and are therefore equal.

Q. E. D.

SEGMENT *of a circle being given, to defcribe the circle of which it is the fegment.*

From any point in the fegment draw —— and —— bifect them, and from the points of bifection draw —— ⊥ —— and —— ⊥ —— where they meet is the centre of the circle.

Becaufe —— terminated in the circle is bifected perpendicularly by ——, it paffes through the centre (B. 3. pr. 1.), likewife —— paffes through the centre, therefore the centre is in the interfection of thefe perpendiculars.

Q. E. D.

 N *equal circles* 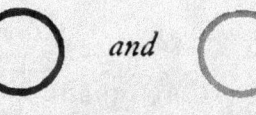 *and* ◯ ,

the arcs ⌣ , ⌣ *on which*

stand equal angles, whether at the centre or circum-

ference, are equal.

First, let 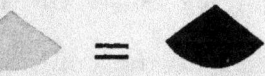 at the centre,

draw ▬▬▬ and ∙∙∙∙∙∙∙ .

Then since ◯ = ◯ ,

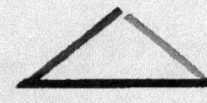 and 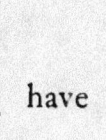 have

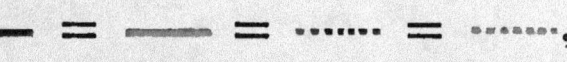 ,

and 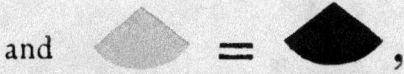 ,

∴ ▬▬▬ = ∙∙∙∙∙∙∙ (B. 1. pr. 4.).

But 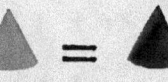 (B. 3. pr. 20.);

∴ ⌒ and ⌒ are similar (B. 3. def. 10.);

they are also equal (B. 3. pr. 24.)

If therefore the equal ſegments be taken from the equal circles, the remaining ſegments will be equal;

hence ⌣ = ⌣ (ax. 3.);

and ∴ ⌣ = ⌣.

But if the given equal angles be at the circumference, it is evident that the angles at the centre, being double of thoſe at the circumference, are alſo equal, and therefore the arcs on which they ſtand are equal.

Q. E. D.

N *equal circles,*  *an*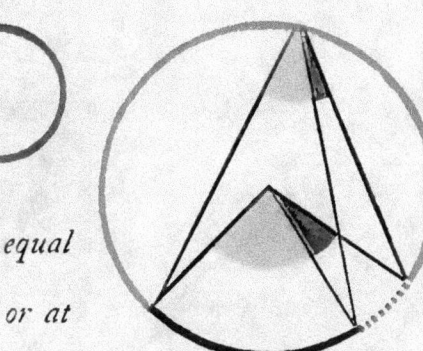

the angles *and*  *which ſtand upon equal arches are equal, whether they be at the centres or at the circumferences.*

For if it be poſſible, let one of them

be greater than the other

and make

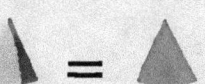

 (B. 3. pr. 26.)

but 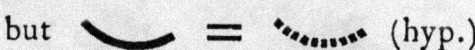 (hyp.)

∴  a part equal

to the whole, which is abſurd; ∴ neither angle

is greater than the other, and

∴ they are equal.

Q. E. D.

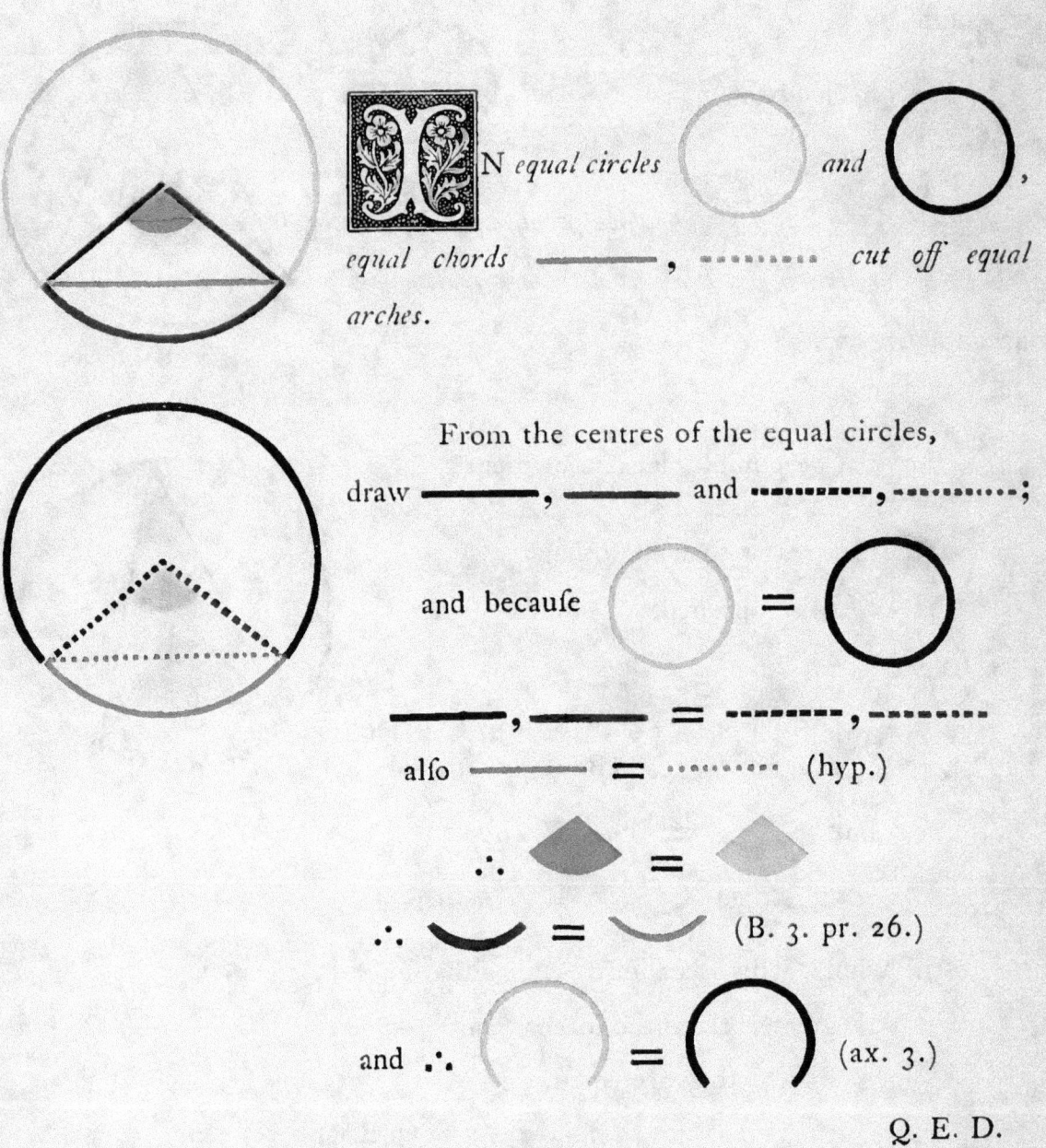

N *equal circles* 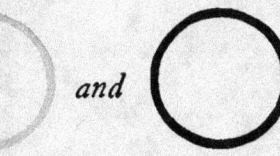 *and*

the chords —————— *and* ⋯⋯⋯⋯ *which sub-*
tend equal arcs are equal.

If the equal arcs be femicircles the propofition is

evident.   But if not,

let ——————, ——————, *and* ▬▬▬▬▬, ⋯⋯⋯⋯

be drawn to the centres ;

becaufe ‿ = ‿ (hyp.)

and 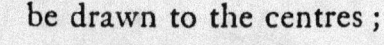 = 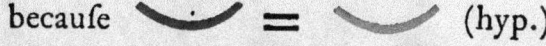 (B. 3. pr. 27.) ;

but —————— *and* —————— = ▬▬▬▬▬ *and* ⋯⋯⋯⋯

∴ —————— = ⋯⋯⋯⋯ (B. 1. pr. 4.) ;

but thefe are the chords fubtending

the equal arcs.

Q. E. D.

Q

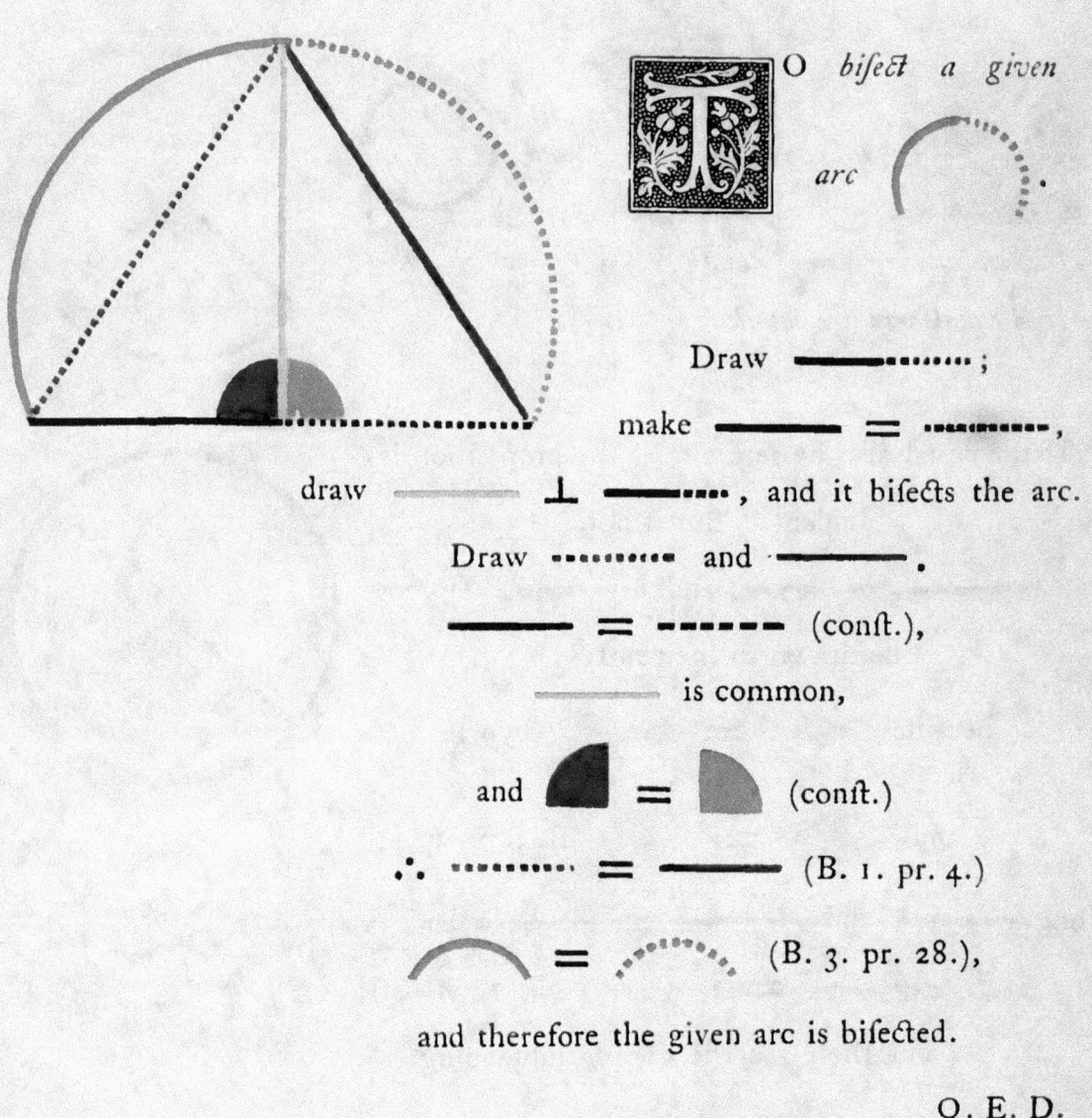

O *bifect   a   given*

arc

Draw ━━━━ ·········· ;

make ━━━━ = ·········· ,

draw ━━━━ ⊥ ━━━··· , and it bifects the arc.

Draw ·········· and ·───── .

━━━━ = ------- (conft.),

━━━━ is common,

and ◗ = ◖ (conft.)

∴ ·········· = ━━━ (B. 1. pr. 4.)

⌒ = ⌒ (B. 3. pr. 28.),

and therefore the given arc is bifected.

Q. E. D.

 N *a circle the angle in a femicircle is a right angle, the angle in a fegment greater than a femicircle is acute, and the angle in a fegment lefs than a femicircle is obtufe.*

FIGURE I.

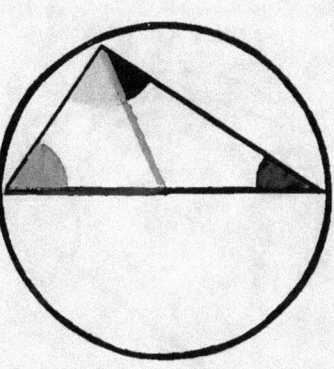

FIGURE I.

The angle in a femicircle is a right angle.

Draw ━━━━ and ━━━━

 and (B. 1. pr. 5.)

 = the half of two

right angles ═ a right angle. (B. 1. pr. 32.)

FIGURE II.

The angle ▲ in a fegment greater than a femi-circle is acute.

FIGURE II.

Draw ━━━ the diameter, and ━━━━

∴ ◣ ═ a right angle

∴ ▲ is acute.

FIGURE III.

FIGURE III.

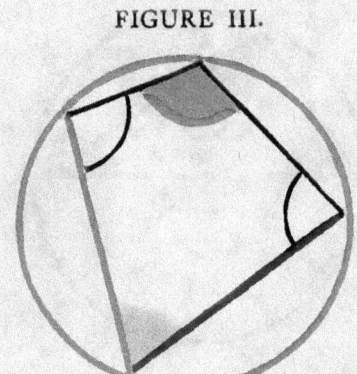

The angle 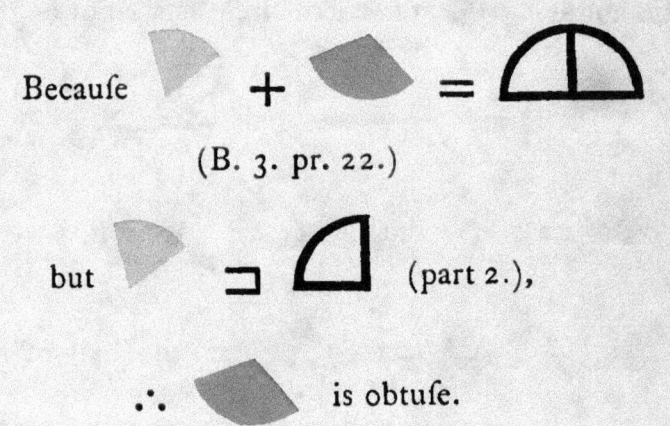 in a segment less than semi-circle is obtuse.

Take in the oppofite circumference any point, to which draw ▬▬▬ and ▬▬▬ .

Becaufe ■ **+** ◗ **=** ⬭

(B. 3. pr. 22.)

but ◖ ⊐ ◫ (part 2.),

∴ ◗ is obtufe.

Q. E. D.

116

 F *a right line* ━━━ *be a tangent to a circle, and from the point of contact a right line* ━━━ *be drawn cutting the circle, the angle*

 made by this line with the tangent

is equal to the angle 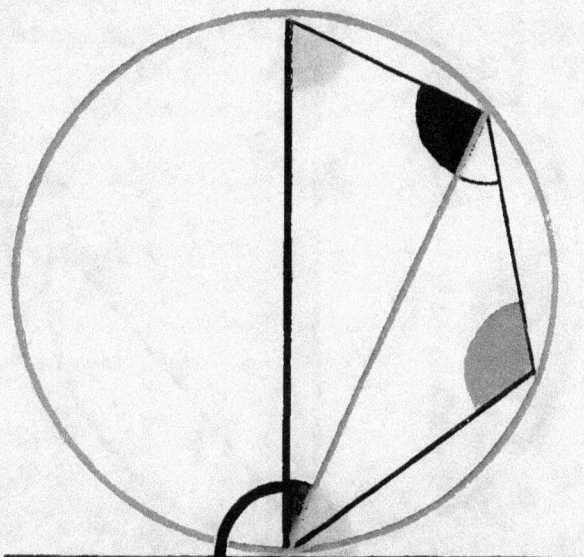 *in the alter- ate segment of the circle.*

If the chord fhould pafs through the centre, it is evi- dent the angles are equal, for each of them is a right angle. (B. 3. prs. 16, 31.)

But if not, draw ━━━ ⊥ ━━━ from the point of contact, it muft pafs through the centre of the circle, (B. 3. pr. 19.)

$$\therefore \blacksquare = \square \quad \text{(B. 3. pr. 31.)}$$

$$\square + \blacksquare = \square = \blacksquare \quad \text{(B. 1. pr. 32.)}$$

$$\therefore \square = \square \quad \text{(ax.)}$$

Again $\square = \square = \square + \square$
(B. 3. pr. 22.)

$$\therefore \square = \square \quad \text{, (ax.), which is the angle in}$$

the alternate fegment.

Q. E. D.

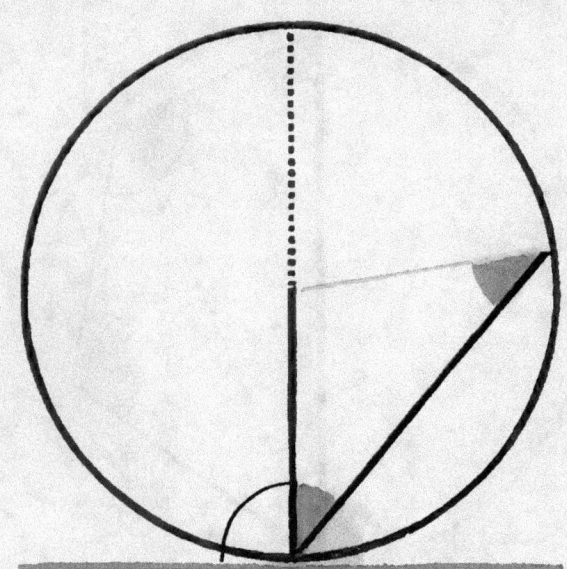

*O*N *a given straight line* ——— *to describe a segment of a circle that shall contain an angle equal to a given angle*

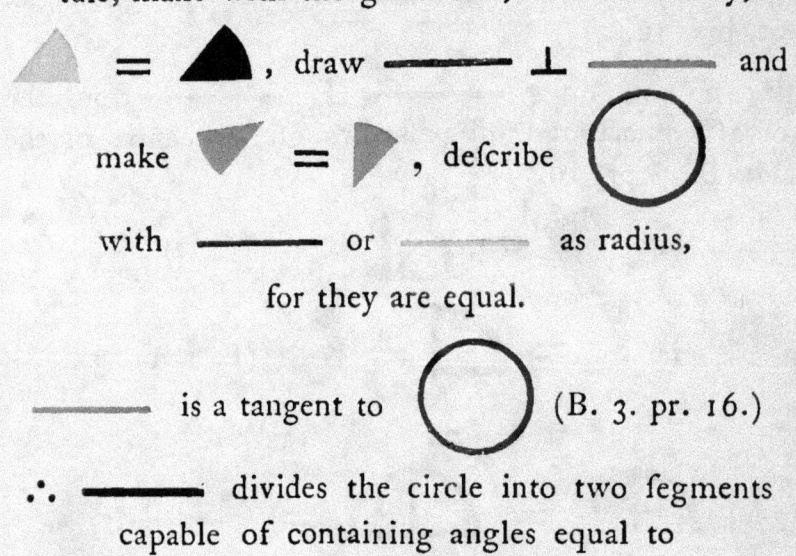

If the given angle be a right angle, bisect the given line, and describe a semicircle on it, this will evidently contain a right angle. (B. 3. pr. 31.)

If the given angle be acute or ob-tufe, make with the given line, at its extremity,

△ = ▲ , draw ——— ⊥ ——— and

make ▽ = ◤ , defcribe ◯

with ——— or ——— as radius, for they are equal.

——— is a tangent to ◯ (B. 3. pr. 16.)

∴ ——— divides the circle into two fegments capable of containing angles equal to

◣ and △ which were made refpectively equal

to ◠ and ▲ (B. 3. pr. 32.)

Q. E. D.

 O cut off from a given cir-

cle 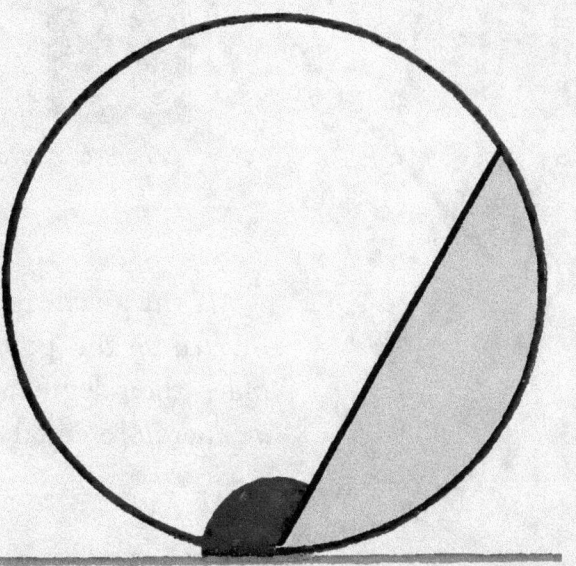 a *fegment*

which *fhall contain an angle equal to a*

given *angle*  .

Draw ———— (B. 3. pr. 17.),

a tangent to the circle at any point;

at the point of contact make

the given angle;

and 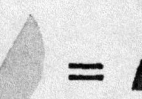 contains an angle ═ the given angle.

Becaufe ———— is a tangent,

and ———— cuts it, the

angle in ═ (B. 3. pr. 32.),

but ═ (conft.)

Q. E. D.

FIGURE I.

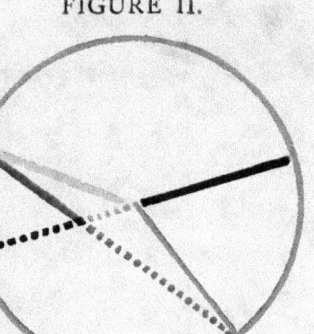

 F *two chords* { ⎯⎯ ⋯⋯ / ▪▪▪▪ ⎯⎯ } *in a circle intersect each other, the rectangle contained by the segments of the one is equal to the rectangle contained by the segments of the other.*

FIGURE I.

If the given right lines paſs through the centre, they are biſected in the point of interſection, hence the rectangles under their ſegments are the ſquares of their halves, and are therefore equal.

FIGURE II.

FIGURE II.

Let ▪▪▪ ⎯⎯ paſs through the centre, and ⎯⎯ ⋯ not; draw ⎯⎯ and ⎯⎯.

Then ⎯⎯ × ⋯⋯ = ⎯⎯^{2} − ⋯^{2} (B. 2. pr. 6.),

or ⎯⎯ × ▪▪▪ = ▪▪▪^{2} − ⋯^{2},

∴ ⎯⎯ × ▪▪▪ = ⎯⎯ × ▪▪▪ (B. 2. pr. 5.).

FIGURE III.

FIGURE III.

Let neither of the given lines paſs through the centre, draw through their interſection a diameter ▪▪▪ ⎯⎯,

and ▪▪▪ × ⎯⎯ = ⎯⎯ × ⋯ (Part. 2.),

also ▪▪▪ × ⎯⎯ = ⎯⎯ × ▪▪▪ (Part. 2.);

∴ ⎯⎯ × ▪▪▪ = ⎯⎯ × ▪▪▪.

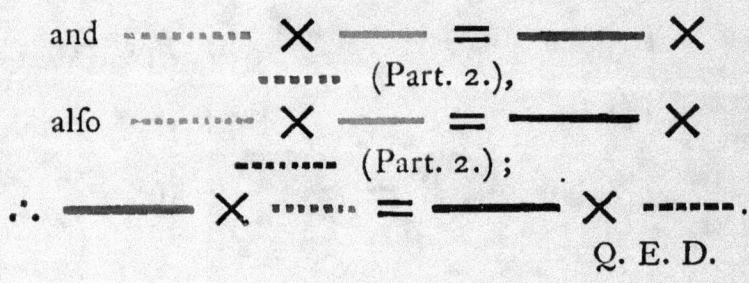

Q. E. D.

F *from a point without a circle two ſtraight lines be drawn to it, one of which* ——— *is a tangent to the circle, and the other* ———·——— *cuts it ; the rectangle under the whole cutting line* ——·——— *and the external ſegment* ——— *is equal to the ſquare of the tangent* ———.

FIGURE I.

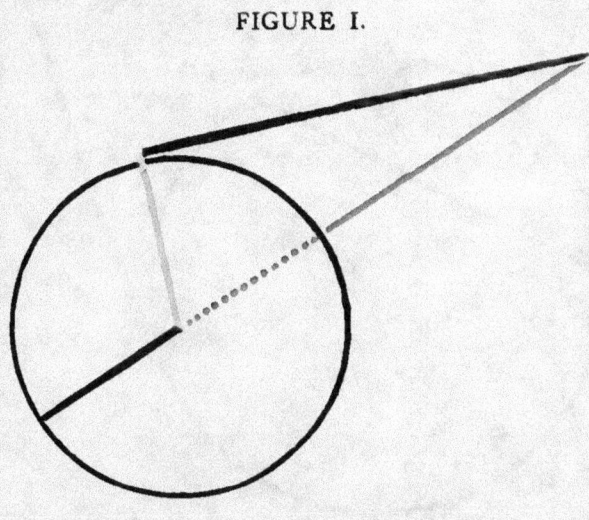

FIGURE I.

Let ——·——— paſs through the centre ;

draw ——— from the centre to the point of contact ;

———^{2} = ——·———^{2} minus ———^{2} (B. 1. pr. 47),

or ———^{2} = ———·———^{2} minus ———·———^{2},

∴ ———^{2} = ——·——— × ——— (B. 2. pr. 6).

FIGURE II.

If ·······——— do not paſs through the centre, draw ———·——— and ·····—.

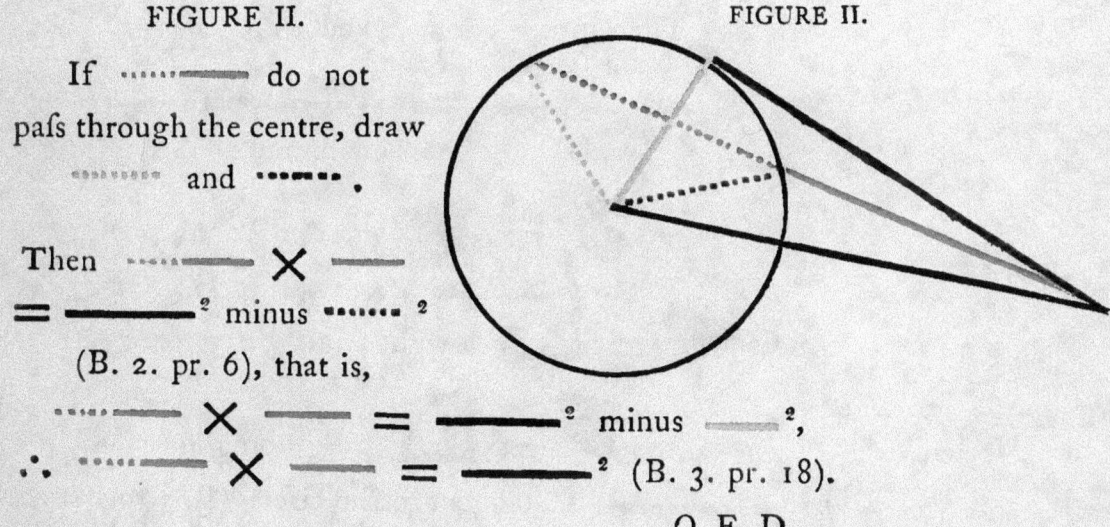

Then ——·——— × ——— = ———^{2} minus ·····—^{2} (B. 2. pr. 6), that is,

——·——— × ——— = ———^{2} minus ———^{2},

∴ ——·——— × ——— = ———^{2} (B. 3. pr. 18).

Q. E. D.

R

121

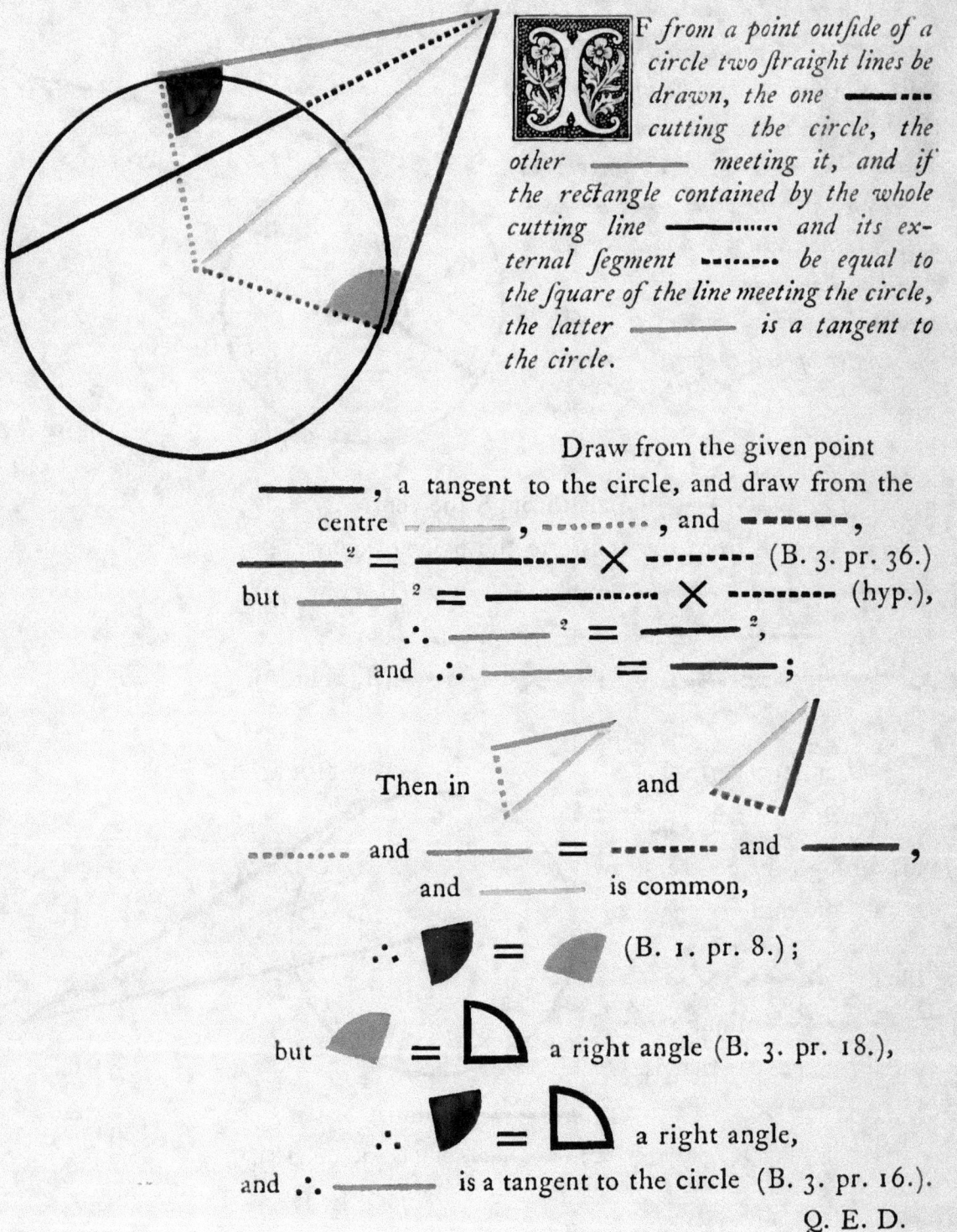

I F *from a point outfide of a circle two ftraight lines be drawn, the one* ▬▬··· *cutting the circle, the other* ▬▬▬ *meeting it, and if the reƈtangle contained by the whole cutting line* ▬▬······ *and its external fegment* ▬······· *be equal to the fquare of the line meeting the circle, the latter* ▬▬▬ *is a tangent to the circle.*

Draw from the given point ▬▬▬, a tangent to the circle, and draw from the centre ▬▬▬, ▬·········, and ▬▬▬▬,

$$▬▬^2 = ▬▬▬······ \times ▬········ \quad (\text{B. 3. pr. 36.})$$

but $▬▬^2 = ▬▬▬······ \times ▬········ \quad (\text{hyp.}),$

$$\therefore ▬▬^2 = ▬▬^2,$$

and $\therefore ▬▬▬ = ▬▬▬ ;$

Then in ◢ and ◣

▬········ and ▬▬▬ = ▬······· and ▬▬▬,

and ▬▬▬ is common,

$$\therefore \; ◥ = ◢ \quad (\text{B. 1. pr. 8.});$$

but $◢ = \llcorner$ a right angle (B. 3. pr. 18.),

$$\therefore \; ◥ = \llcorner \text{ a right angle,}$$

and $\therefore ▬▬▬$ is a tangent to the circle (B. 3. pr. 16.).

Q. E. D.

BOOK IV.

DEFINITIONS.

I.

 RECTILINEAR figure is said to be *inscribed in* another, when all the angular points of the inscribed figure are on the sides of the figure in which it is said to be inscribed.

II.

A FIGURE is said to be *described about* another figure, when all the sides of the circumscribed figure pass through the angular points of the other figure.

III.

A RECTILINEAR figure is said to be *inscribed in* a circle, when the vertex of each angle of the figure is in the circumference of the circle.

IV.

A RECTILINEAR figure is said to be *circumscribed about* a circle, when each of its sides is a tangent to the circle.

V.

A CIRCLE is said to be *inscribed in* a rectilinear figure, when each side of the figure is a tangent to the circle.

VI.

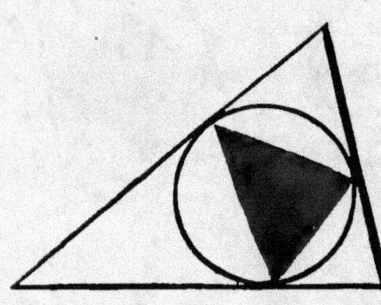

A CIRCLE is said to be *circumscribed about* a rectilinear figure, when the circumference passes through the vertex of each angle of the figure.

 is circumscribed.

VII.

A STRAIGHT line is said to be *inscribed in* a circle, when its extremities are in the circumference.

The Fourth Book of the Elements is devoted to the solution of problems, chiefly relating to the inscription and circumscription of regular polygons and circles.

A regular polygon is one whose angles and sides are equal.

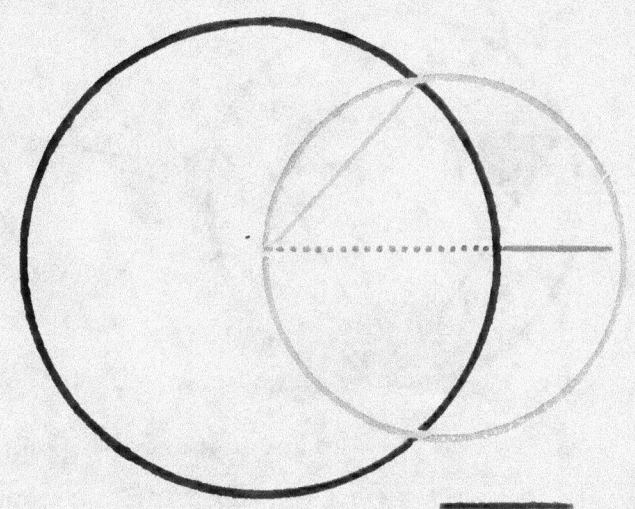

 N *a given circle* ◯ *to place a straight line, equal to a given straight line* (———), *not greater than the diameter of the circle.*

Draw ┅┅━ , the diameter of ◯ ;

and if ┅┅━ = ———— , then

the problem is solved.

But if ┅┅━ be not equal to ———— ,

┅┅━ ⊏ ———— (hyp.);

make ┅┅━ = ———— (B. 1. pr. 3.) with

━ ━ ━ as radius,

describe ◯ , cutting ◯ , and

draw ————, which is the line required.

For ———— = ┅┅━ = ————

(B. 1. def. 15. const.)

Q. E. D.

N *a given circle*

to in-

scribe a triangle equiangular
to a given triangle.

To any point of the given circle draw ———, a tangent (B. 3. pr. 17.); and at the point of contact

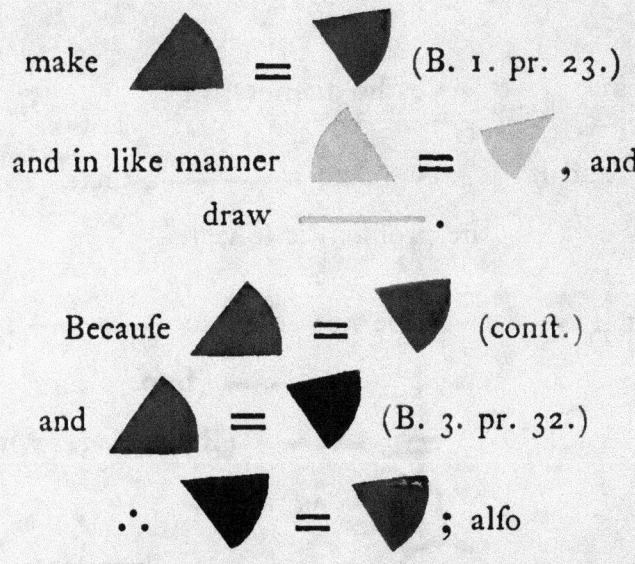

make ▲ = ◣ (B. 1. pr. 23.)

and in like manner ◤ = ◥ , and

draw ——— .

Because ◣ = ◢ (conſt.)

and ◣ = ◥ (B. 3. pr. 32.)

∴ ◢ = ◣ ; alſo

▽ = ◥ for the ſame reaſon.

∴ ◥ = ◥ (B. 1. pr. 32.),

and therefore the triangle inſcribed in the circle is equiangular to the given one.

Q. E. D.

BOUT *a given circle* ⬭ *to circumſcribe a triangle equiangular to a given triangle.*

Produce any ſide ——————, of the given triangle both ways; from the centre of the given circle draw ——————, any radius.

Make ◗ = ◗ (B. 1. pr. 23.)

and ◗ = ◗ .

At the extremities of the three radii, draw ——————, and —————— and ┄┄┄┄┄, tangents to the given circle. (B. 3. pr. 17.)

The four angles of ◣ , taken together, are equal to four right angles. (B. 1. pr. 32.)

but 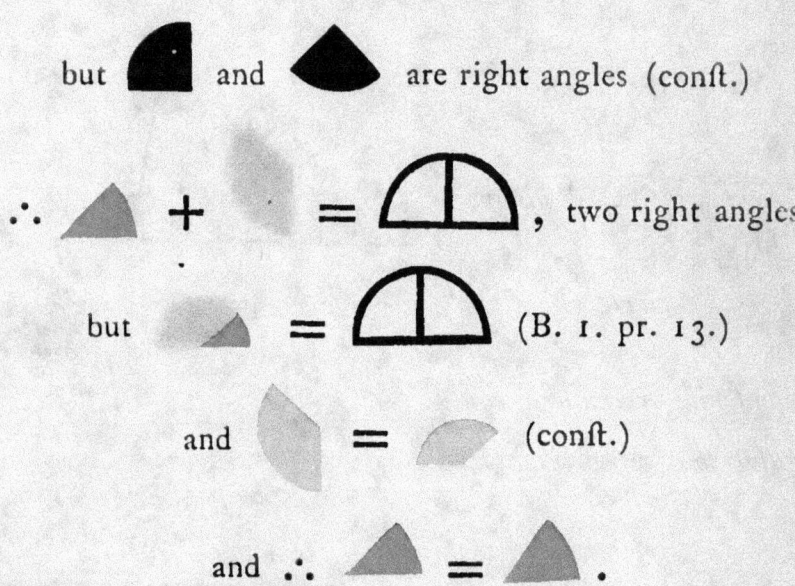 and ▰ are right angles (conſt.)

∴ ▲ + ◖ = ⬭ , two right angles

but ◣ = ⬭ (B. 1. pr. 13.)

and ◗ = ◗ (conſt.)

and ∴ ▲ = ▲ .

In the ſame manner it can be demonſtrated that

◺ = ◺ ;

∴ ◢ = ◢ (B. 1. pr. 32.)

and therefore the triangle circumſcribed about the given circle is equiangular to the given triangle.

Q. E. D.

 N *a given triangle* 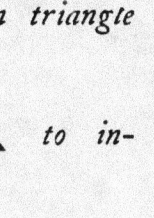 *to in-*
scribe a circle.

Bifect 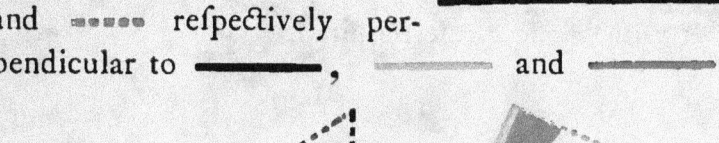 and (B. 1. pr. 9.) by ⋯⋯⋯
and ▬▬▬ ;
from the point where thefe lines
meet draw ▬▬▬ ,
and ▬▬▬ refpectively per-
pendicular to ▬▬▬ , ▬▬▬ and ▬▬▬ .

In 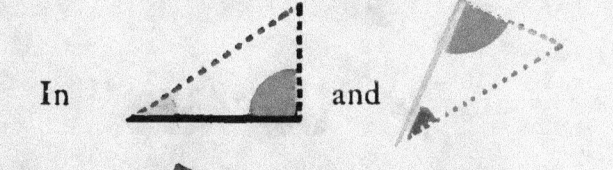 and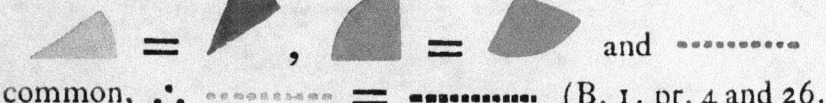

▲ = ◤ , ◗ = ◗ and ⋯⋯⋯
common, ∴ ▬▬▬ = ▬▬▬ (B. 1. pr. 4 and 26.)

In like manner, it may be fhown alfo
that ▬▬▬ = ▬▬▬ ,
∴ ▬▬▬ = ▬▬▬ = ▬▬▬ ;
hence with any one of thefe lines as radius, defcribe

○ and it will pafs through the extremities of the

other two; and the fides of the given triangle, being perpendicular to the three radii at their extremities, touch the
circle (B. 3. pr. 16.), which is therefore infcribed in the
given circle.

s Q. E. D.

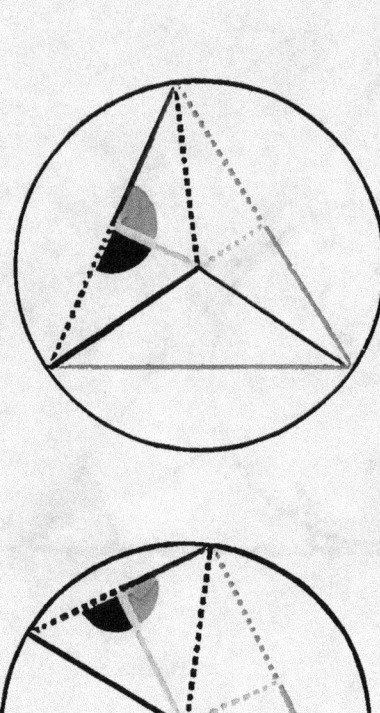

 O *defcribe a circle about a given triangle.*

Make ━━━ = ┅┅ and ━━━ =
┅┅ (B. 1. pr. 10.)

From the points of bifection draw ━━━ and
┅┅ ⊥ ━━━ and ━━━ refpec-
tively (B. 1. pr. 11.), and from their point of
concourfe draw ━━━ , ┅┅ and ━━━
and defcribe a circle with any one of them, and
it will be the circle required.

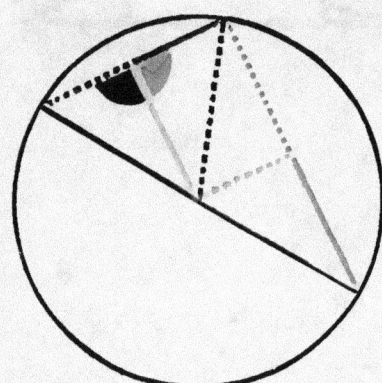

In 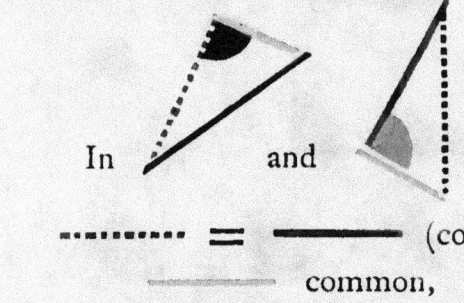 and

┅┅ = ━━━ (conft.),

━━━ common,

◗ = ◗ (conft.),

∴ ━━━ = ┅┅ (B. 1. pr. 4.).

In like manner it may be fhown that

━━━ = ┅┅

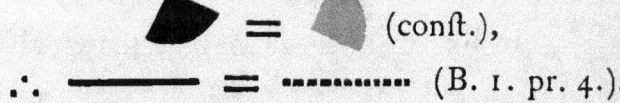

∴ ┅┅ = ━━━ = ━━━ ; and
therefore a circle defcribed from the concourfe of
thefe three lines with any one of them as a radius
will circumfcribe the given triangle.

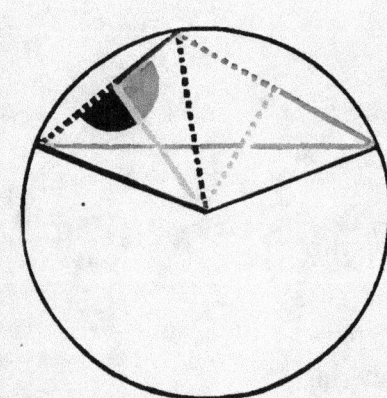

Q. E. D.

N *a given circle* 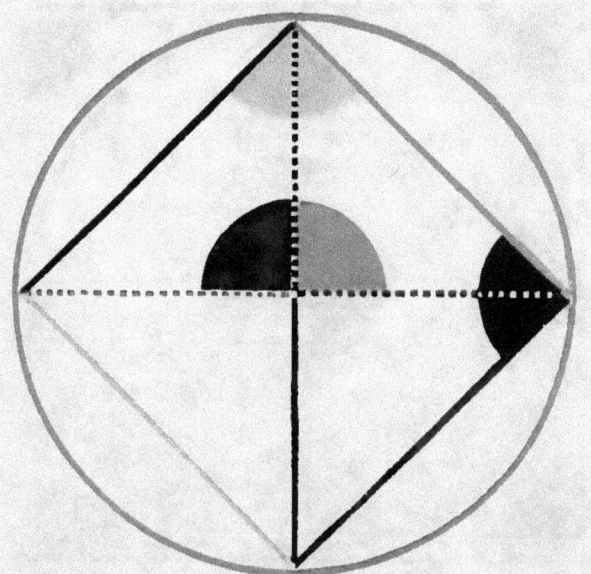 *to*

inscribe a square.

Draw the two diameters of the circle ⊥ to each other, and draw ━━━ , ━━━ , ━━━ and ━━━

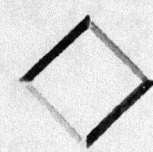 is a square.

For, since  and are, each of them, in a semicircle, they are right angles (B. 3. pr. 31),

∴ ━━━ ‖ ━━━ (B. 1. pr. 28):

and in like manner ━━━ ‖ ━━━ .

And because █ = █ (conft.), and

┄┄┄ = ┄┄┄ = ┄┄┄ (B. 1. def. 15).

∴ ━━━ = ━━━ (B. 1. pr. 4);

and since the adjacent sides and angles of the parallelogram ◇ are equal, they are all equal (B. 1. pr. 34);

and ∴ ◇ , inscribed in the given circle, is a square.

Q. E. D.

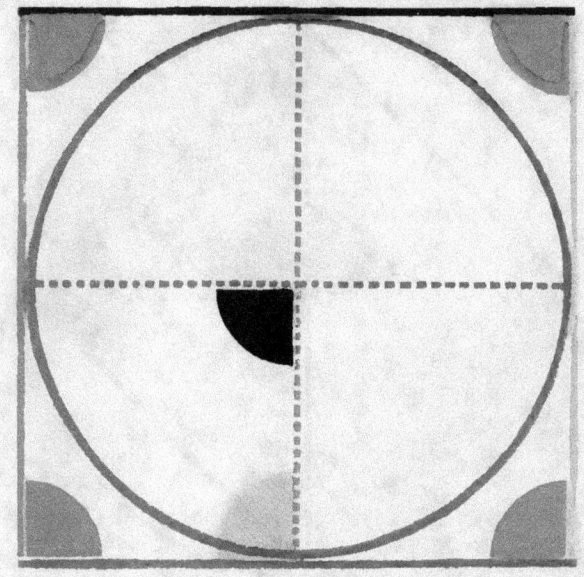

BOUT *a given circle*

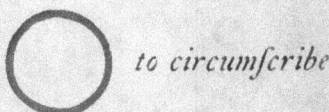 *to circumfcribe*

a fquare.

Draw two diameters of the given circle perpendicular to each other, and through their extremities draw ——, ——, ——, and —— tangents to the circle;

and ▢ is a fquare.

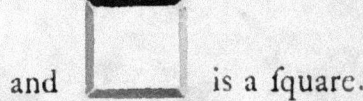 a right angle, (B. 3. pr. 18.)

alfo 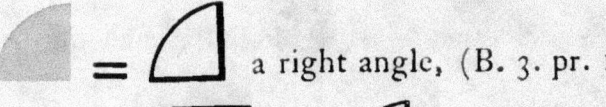 (conft.),

∴ —— ‖ ....; in the fame manner it can be demonftrated that —— ‖ ...., and alfo that —— and —— ‖ ....;

∴ ▢ is a parallelogram, and

because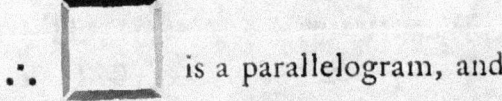

they are all right angles (B. 1. pr. 34):

it is alfo evident that ——, ——, —— and —— are equal.

∴ ▢ is a fquare.

Q. E. D.

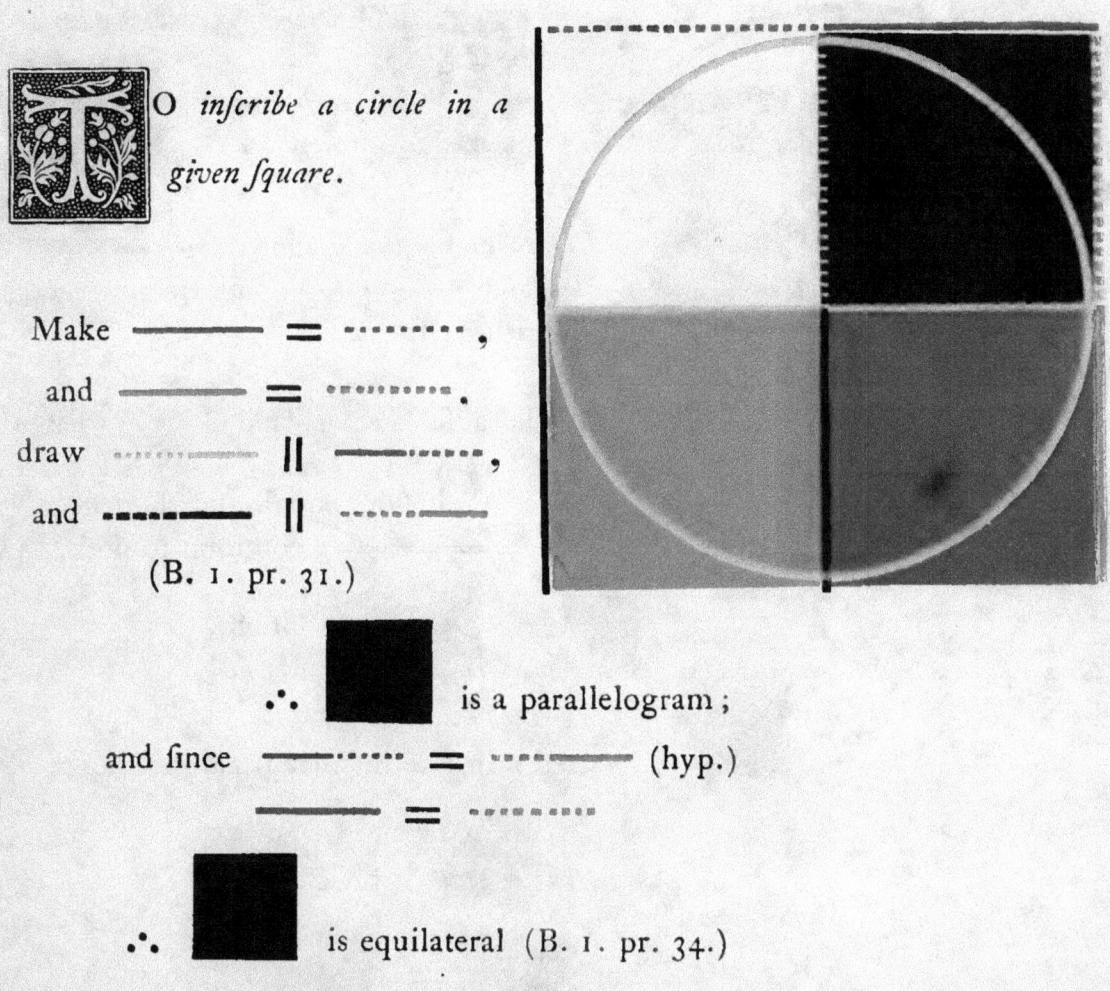

O *inscribe a circle in a given square.*

Make ——— = ········· ,

and ——— = ········· .

draw ········· ‖ ········· ,

and ········· ‖ ·········

(B. 1. pr. 31.)

∴ ▮ is a parallelogram ;

and since ——·····—— = ·····——— (hyp.)

——— = ···········

∴ ▮ is equilateral (B. 1. pr. 34.)

In like manner, it can be shown that

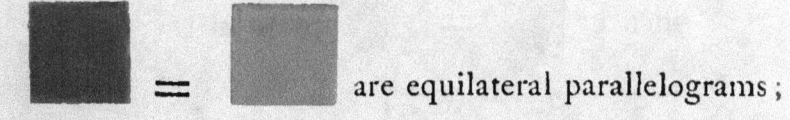

are equilateral parallelograms ;

∴ ·······— = ·······— = ——— = ——— .

and therefore if a circle be described from the concourse of these lines with any one of them as radius, it will be inscribed in the given square. (B. 3. pr. 16.)

Q. E. D.

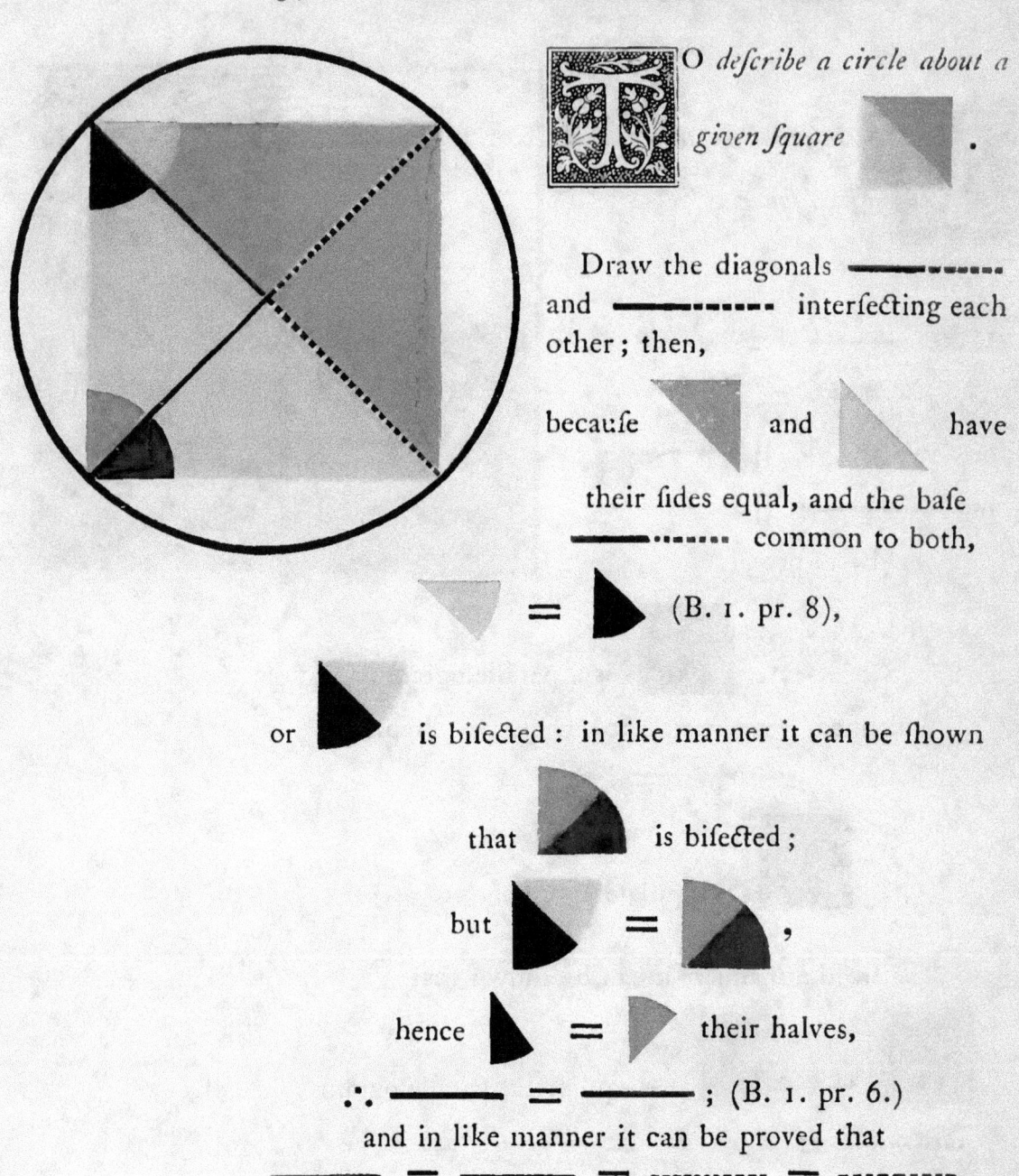

T O *defcribe a circle about a given fquare* .

Draw the diagonals ━━━ ┄┄┄ and ━━━ ┄┄┄ interfecting each other; then,

becaufe and have

their fides equal, and the bafe ━━━ ┄┄┄ common to both,

= (B. 1. pr. 8),

or is bifected : in like manner it can be fhown

that is bifected ;

but = ,

hence = their halves,

∴ ━━━ = ━━━ ; (B. 1. pr. 6.)

and in like manner it can be proved that

━━━ = ━━━ = ┅┅┅ = ┄┄┄ .

If from the confluence of thefe lines with any one of them as radius, a circle be defcribed, it will circumfcribe the given fquare.

Q. E. D.

O *construct an isosceles triangle, in which each of the angles at the base shall be double of the vertical angle.*

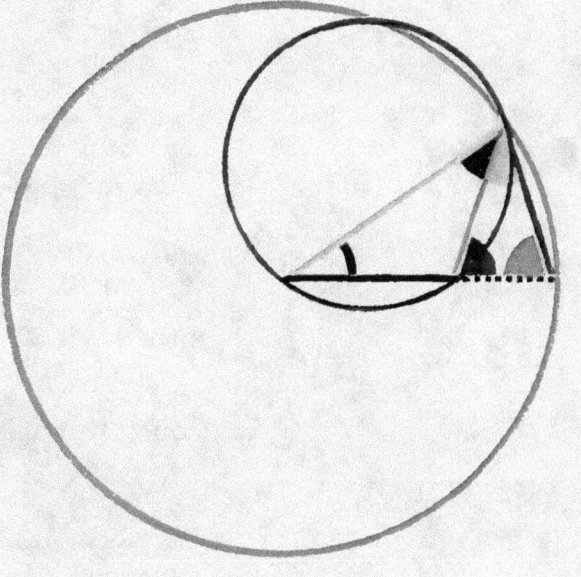

Take any straight line ━━━⋯⋯ and divide it so that

━━⋯ ✕ ⋯ = ━━━ ²

(B. 2. pr. 11.)

With ━━━⋯ as radius, describe ⬭ and place

in it from the extremity of the radius, ━━ = ━━,

(B. 4. pr. 1); draw ━━ .

Then 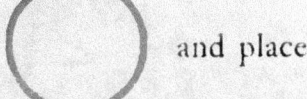 is the required triangle.

For, draw ━━ and describe

◯ about ◺ (B. 4. pr. 5.)

Since ━━⋯ ✕ ⋯ = ━━ ² = ━━ ²,

∴ ━━ is a tangent to ◯ (B. 3. pr. 37.)

∴ ◢ = ◸ (B. 3. pr. 32),

add ◀ to each,

∴ ◭ + ◀ = △ + ◀ ;

but ◀ + ◭ or ◀ = ◗ (B. 1. pr. 5) :

since ——— = ——••••• (B. 1. pr. 5.)

consequently ◗ = △ + ◀ = ◗

(B. 1. pr. 32.)

∴ ——— = ——— (B. 1. pr. 6.)

∴ ——— = ——— = ——— (conſt.)

∴ △ = ◀ (B. 1. pr. 5.)

∴ ◗ = ◀ = ◗ = △ +

◀ = twice △ ; and conſequently each angle at
the baſe is double of the vertical angle.

Q. E. D.

  N *a given circle*

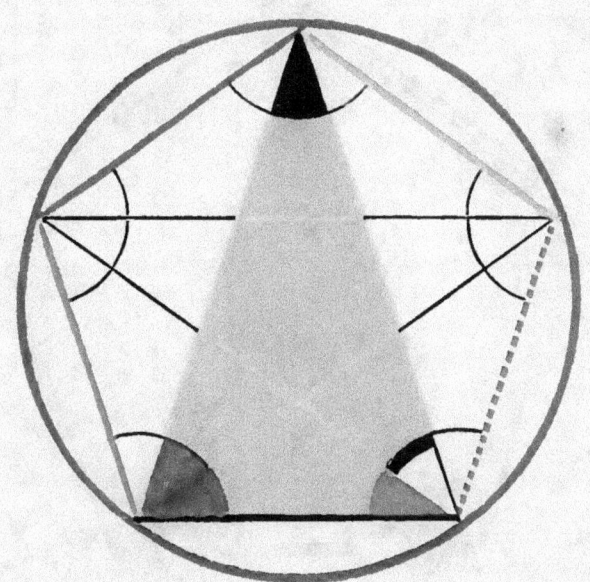

to inscribe an equilateral and equi-angular pentagon.

Construct an isosceles triangle, in which each of the angles at the base shall be double of the angle at the vertex, and inscribe in the given

circle a triangle  equiangular to it; (B. 4. pr. 2.)

Bisect  and 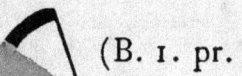 (B. 1. pr. 9.)

draw ———— , ———— . ———— and -------- .

Because each of the angles

 ,  , 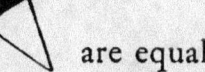 and are equal,

the arcs upon which they stand are equal, (B. 3. pr. 26.) and ∴ ————, ————, ————, ———— and -------- which subtend these arcs are equal (B. 3. pr. 29.) and ∴ the pentagon is equilateral, it is also equiangular, as each of its angles stand upon equal arcs. (B. 3. pr. 27).

Q. E. D.

T

137

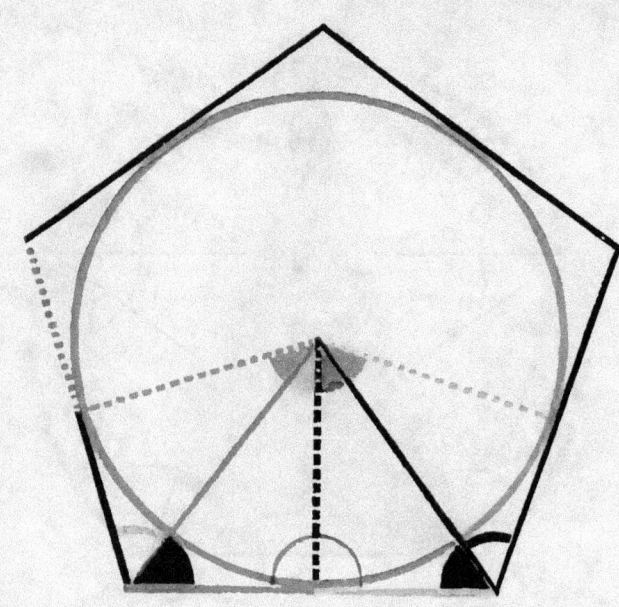

O *defcribe an equilateral and equiangular penta-gon about a given circle*

Draw five tangents through the vertices of the angles of any regular pentagon infcribed in the given circle (B. 3. pr. 17).

Thefe five tangents will form the required pentagon.

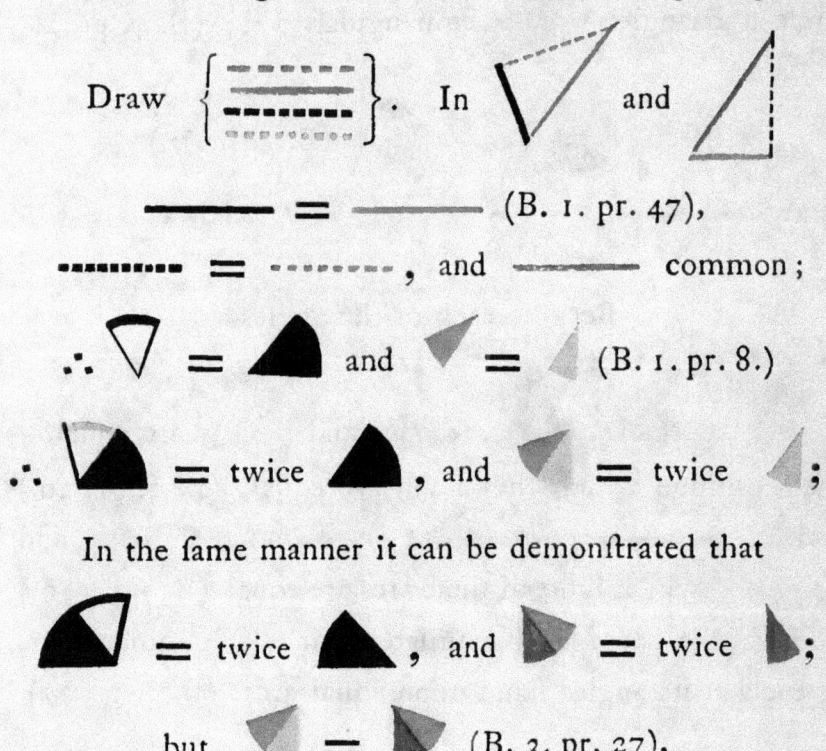

∴ their halves ◤ = ◢, alſo ◖ = ◗, and

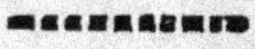

⚊⚊⚊ common;

∴ ▲ = ▲ and ━━━ = ━━━ ,

∴ ━━━ = twice ━━ ;

In the ſame manner it can be demonſtrated

that ━━━━━ = twice ━━ ,

but ━━ = ━━

∴ ━━━━━ = ━━ ━━ ;

In the ſame manner it can be demonſtrated that the
other ſides are equal, and therefore the pentagon is equilateral, it is alſo equiangular, for

◣ = twice ▲ and ◤ = twice ▲ ,

and therefore ▲ = ▲ ,

∴ ◣ = ◤ ; in the ſame manner it can be
demonſtrated that the other angles of the deſcribed
pentagon are equal.

Q. E. D

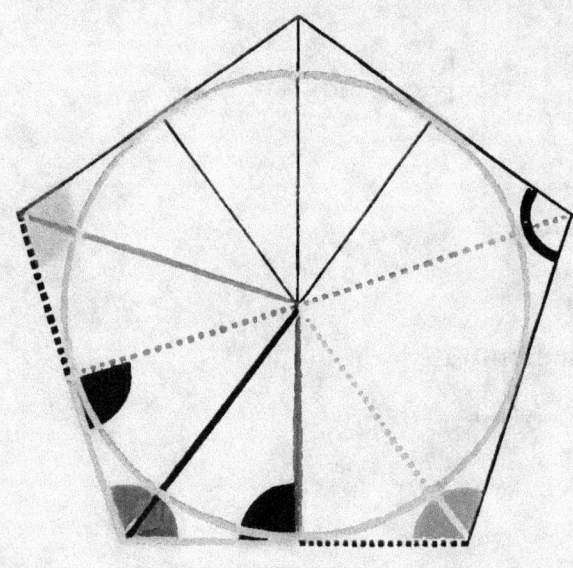

O *inscribe a circle in a given equiangular and equilateral pentagon.*

Let [pentagon] be a given equiangular and equilateral pentagon; it is required to inscribe a circle in it.

Make [▼] = [▲], and [▲] = [▼] (B. 1. pr. 9.)

Draw ━ ━ ━ . ━━━ , ━━━ , ━━━ , &c.

Because ▬▬ = ▬▬ , [▼] = [▲],

and ━━━ common to the two triangles

and ;

∴ ━━━ = ━━━ and = [▲] (B. 1. pr. 4.)

And because = = twice [▲]

∴ = twice [▼] , hence is bisected by ━━━ .

In like manner it may be demonstrated that [◁] is

bisected by ━━━ , and that the remaining angle of the polygon is bisected in a similar manner.

140

Draw ——— , ·········· , &c. perpendicular to the sides of the pentagon.

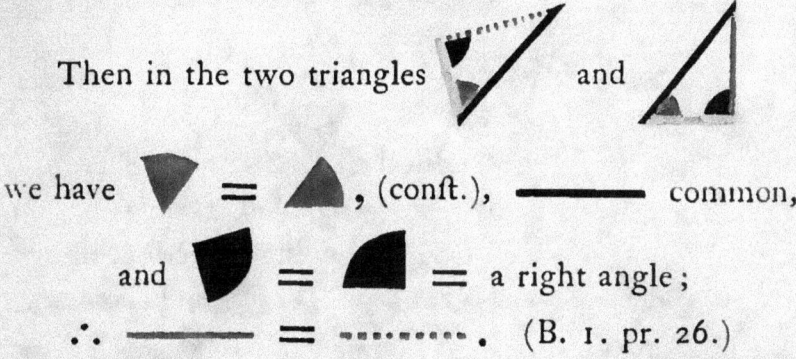

Then in the two triangles   and

we have ▼ = ▲ , (conft.), ——— common,

and ◣ = ◤ = a right angle;

∴ ——— = ·········· . (B. 1. pr. 26.)

In the fame way it may be fhown that the five perpendiculars on the fides of the pentagon are equal to one another.

Defcribe ◯ with any one of the perpendiculars as radius, and it will be the infcribed circle required. For if it does not touch the fides of the pentagon, but cut them, then a line drawn from the extremity at right angles to the diameter of a circle will fall within the circle, which has been fhown to be abfurd. (B. 3. pr. 16.)

Q. E. D.

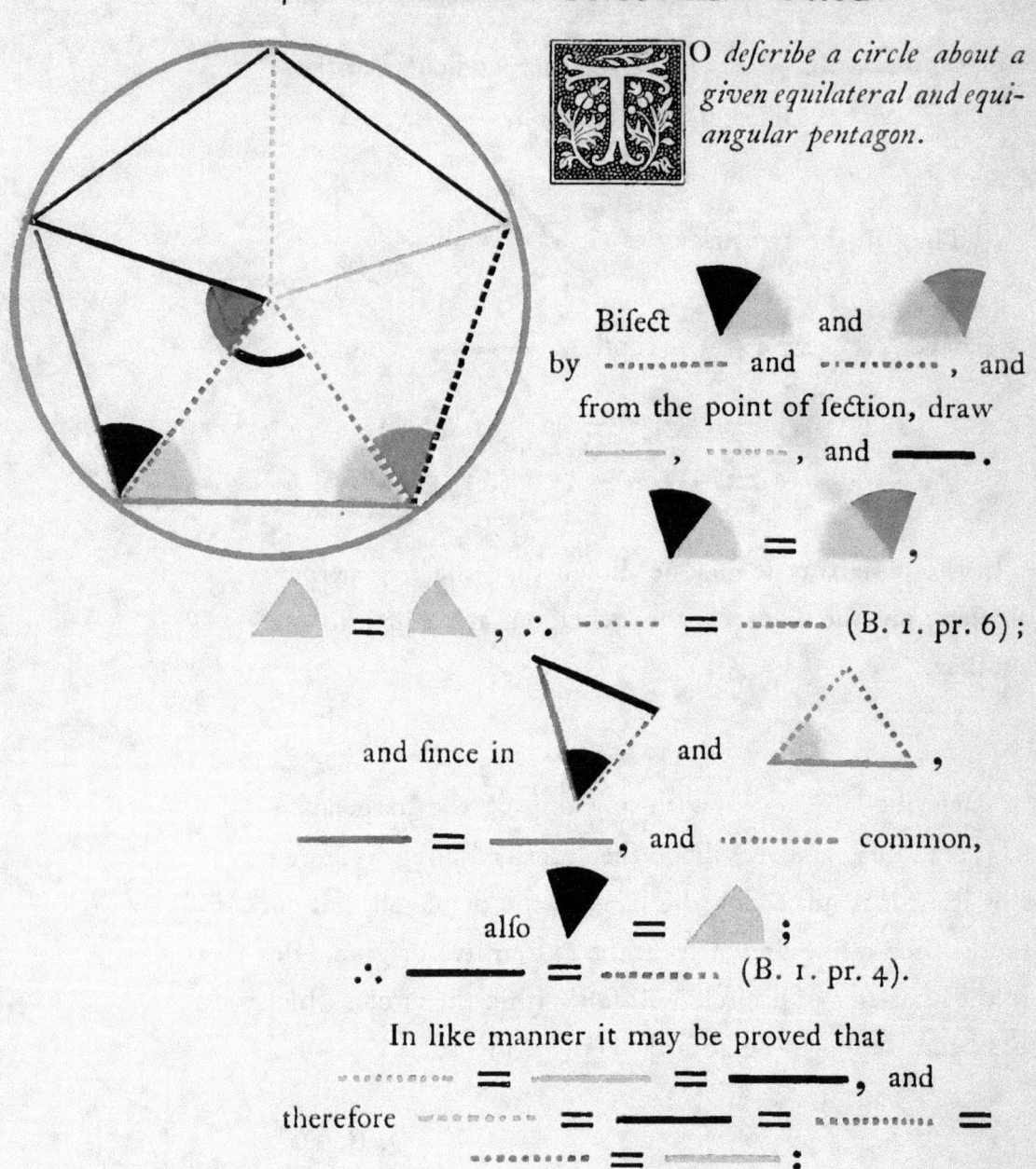

O defcribe a circle about a given equilateral and equi-angular pentagon.

Bifect ◤ and ◥, and by ·········· and ·-·-·-·-· , and from the point of fection, draw ━━━ , ·-·-·-· , and ━━━ .

◤ = ◥ ,

▲ = ▲ , ∴ ·-·-·- = ·-·-·-· (B. 1. pr. 6);

and fince in ◺ and △ ,

━━━ = ━━━ , and ·········· common,

alfo ◤ = ▲ ;

∴ ━━━ = ·-·-·-· (B. 1. pr. 4).

In like manner it may be proved that

·-·-·-· = ·-·-·-· = ━━━ , and

therefore ·-·-·-· = ━━━ = ·-·-·-· = ·-·-·-· = ━━━ :

Therefore if a circle be defcribed from the point where thefe five lines meet, with any one of them as a radius, it will circumfcribe the given pentagon.

Q. E. D.

O *inſcribe an equilateral and equian-gular hexagon in a given circle*

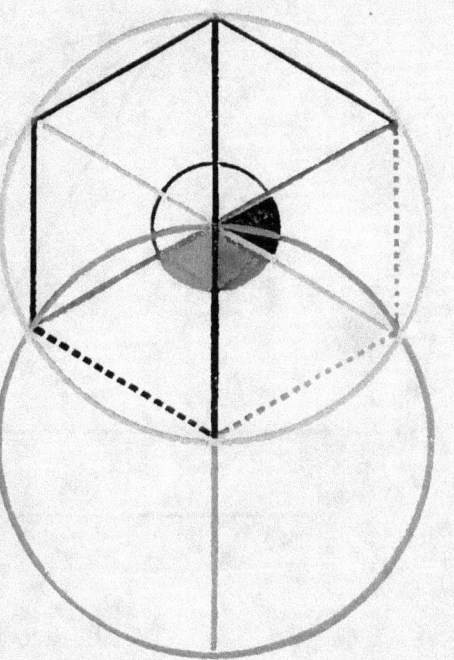

From any point in the circumference of

the given circle deſcribe ◯ paſſing

through its centre, and draw the diameters

━━━ , ━━━ and ─── ; draw

▬▬▬ , ┄┄┄ , ········· , &c. and the required hexagon is inſcribed in the given circle.

Since ━━━ paſſes through the centres

of the circles, ◺ and ◿ are equilateral

triangles, hence ◗ = ◗ = one-third of two right

angles ; (B. 1. pr. 32) but ◗ = ◓ (B. 1. pr. 13);

∴ ◀ = ▶ = ◀ = one-third of ◓ (B. 1. pr. 32), and the angles vertically oppoſite to theſe are all equal to one another (B. 1. pr. 15), and ſtand on equal arches (B. 3. pr. 26), which are ſubtended by equal chords (B. 3. pr. 29) ; and ſince each of the angles of the hexagon is double of the angle of an equilateral triangle, it is alſo equiangular.   Q. E. D.

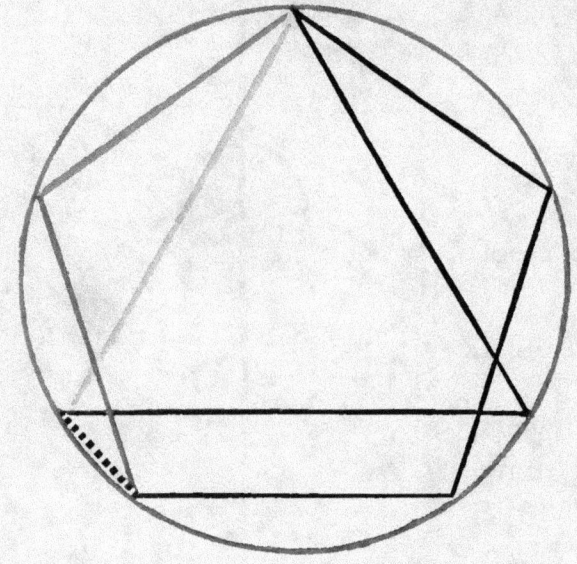

O *inscribe an equilateral and equiangular quindecagon in a given circle.*

Let ———— and ———— be the sides of an equilateral pentagon inscribed in the given circle, and ———— the side of an inscribed equilateral triangle.

The arc subtended by
———— and ———— $\Big\} = \frac{2}{5} = \frac{6}{15} \Big\{$ of the whole circumference.

The arc subtended by
———— $\Big\} = \frac{1}{3} = \frac{5}{15} \Big\{$ of the whole circumference.

Their difference $= \frac{1}{15}$

∴ the arc subtended by ·········· $= \frac{1}{15}$ difference of the whole circumference.

Hence if straight lines equal to ·········· be placed in the circle (B. 4. pr. 1), an equilateral and equiangular quindecagon will be thus inscribed in the circle.

Q. E. D.

BOOK V.

DEFINITIONS.

I.

 LESS magnitude is faid to be an aliquot part or fubmultiple of a greater magnitude, when the lefs meafures the greater; that is, when the lefs is contained a certain number of times exactly in the greater.

II.

A GREATER magnitude is faid to be a multiple of a lefs, when the greater is meafured by the lefs; that is, when the greater contains the lefs a certain number of times exactly.

III.

RATIO is the relation which one quantity bears to another of the fame kind, with refpect to magnitude.

IV.

MAGNITUDES are faid to have a ratio to one another, when they are of the fame kind; and the one which is not the greater can be multiplied fo as to exceed the other.

The other definitions will be given throughout the book where their aid is firft required.

U

AXIOMS.

I.

QUIMULTIPLES or equisubmultiples of the same, or of equal magnitudes, are equal.

If A $=$ B, then
twice A $=$ twice B, that is,
2 A $=$ 2 B;
3 A $=$ 3 B;
4 A $=$ 4 B;
&c. &c.
and $\frac{1}{2}$ of A $=$ $\frac{1}{2}$ of B;
$\frac{1}{3}$ of A $=$ $\frac{1}{3}$ of B;
&c. &c.

II.

A MULTIPLE of a greater magnitude is greater than the same multiple of a less.

Let A $\sqsubset$ B, then
2 A $\sqsubset$ 2 B;
3 A $\sqsubset$ 3 B;
4 A $\sqsubset$ 4 B;
&c. &c.

III.

THAT magnitude, of which a multiple is greater than the same multiple of another, is greater than the other.

Let 2 A $\sqsubset$ 2 B, then
A $\sqsubset$ B;
or, let 3 A $\sqsubset$ 3 B, then
A $\sqsubset$ B;
or, let m A $\sqsubset$ m B, then
A $\sqsubset$ B.

F *any number of magnitudes be equimultiples of as many others, each of each: what multiple soever any one of the first is of its part, the same multiple shall of the first magnitudes taken together be of all the others taken together.*

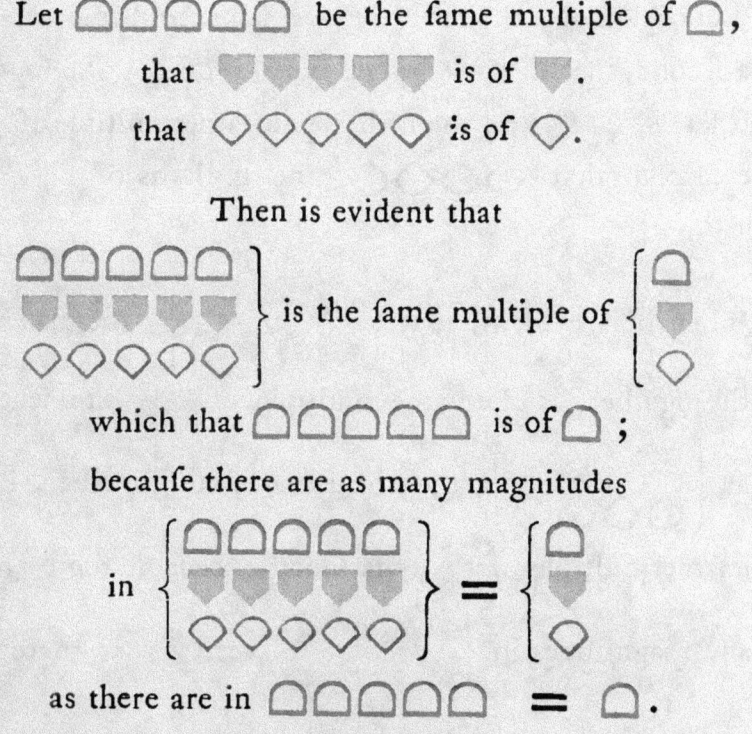

The fame demonftration holds in any number of magnitudes,. which has here been applied to three.

∴ If any number of magnitudes, &c.

F *the firſt magnitude be the ſame multiple of the ſecond that the third is of the fourth, and the fifth the ſame multiple of the ſecond that the ſixth is of the fourth, then ſhall the firſt, together with the fifth, be the ſame multiple of the ſecond that the third, together with the ſixth, is of the fourth.*

Let ⬤⬤⬤, the firſt, be the ſame multiple of ⬤, the ſecond, that ◯◯◯, the third, is of ◯, the fourth; and let ⬤⬤⬤⬤, the fifth, be the ſame multiple of ⬤, the ſecond, that ◯◯◯◯, the ſixth, is of ◯, the fourth.

Then it is evident, that { ⬤⬤⬤ ⬤⬤⬤⬤ }, the firſt and fifth together, is the ſame multiple of ⬤, the ſecond, that { ◯◯◯ ◯◯◯◯ }, the third and ſixth together, is of the ſame multiple of ◯, the fourth; becauſe there are as many magnitudes in { ⬤⬤⬤ ⬤⬤⬤⬤ } = ⬤ as there are in { ◯◯◯ ◯◯◯◯ } = ◯ .

∴ If the firſt magnitude, &c.

 F *the firſt of four magnitudes be the ſame multiple of the ſecond that the third is of the fourth, and if any equimultiples whatever of the firſt and third be taken, thoſe ſhall be equimultiples; one of the ſecond, and the other of the fourth.*

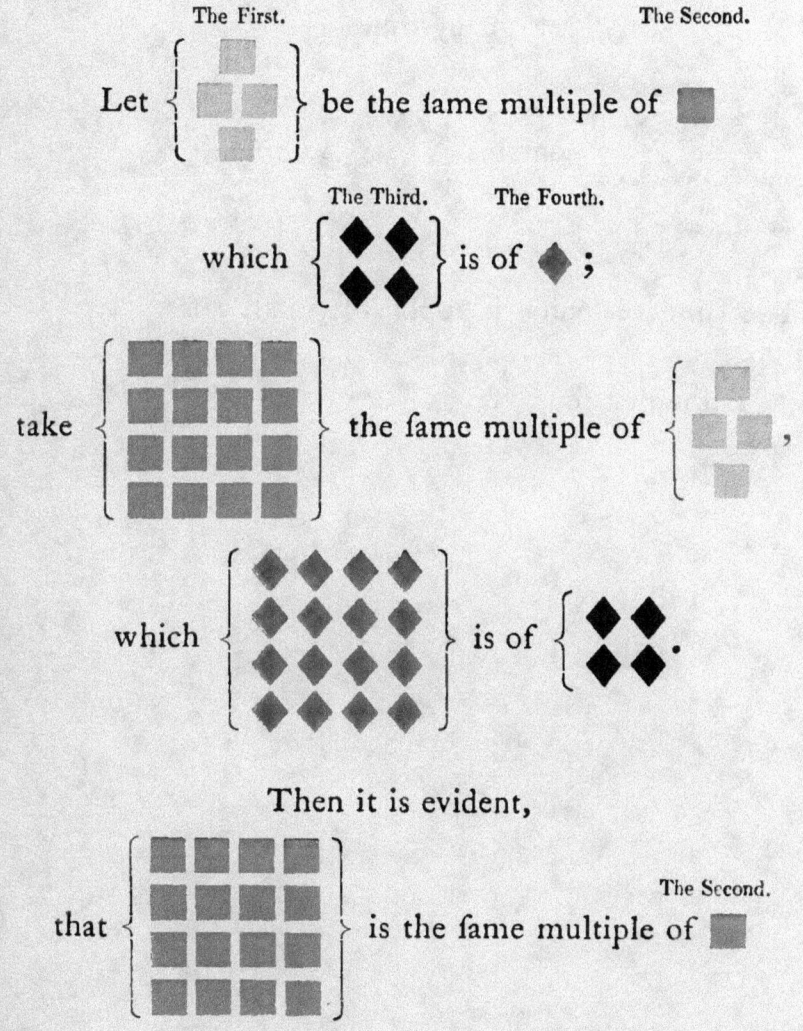

The First.   The Second.

Let { } be the ſame multiple of

The Third.   The Fourth.

which { } is of ;

take { } the ſame multiple of { },

which { } is of { }.

Then it is evident,

The Second.

that { } is the ſame multiple of

which 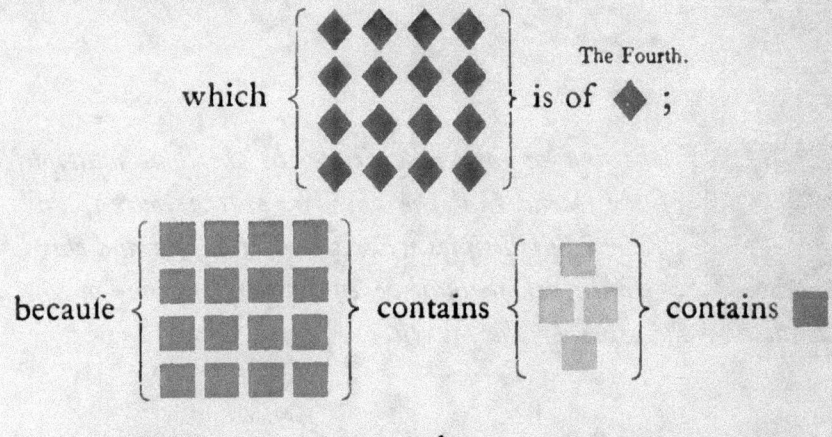 is of ◆ ;

because  contains ▪ contains ▪

as many times as

contains ◆◆ contains ◆ .

The same reasoning is applicable in all cases.

∴ If the first four, &c.

DEFINITION V.

FOUR magnitudes, ●, ▢, ◆, ▼, are said to be proportionals when every equimultiple of the firſt and third be
taken, and every equimultiple of the ſecond and fourth, as,

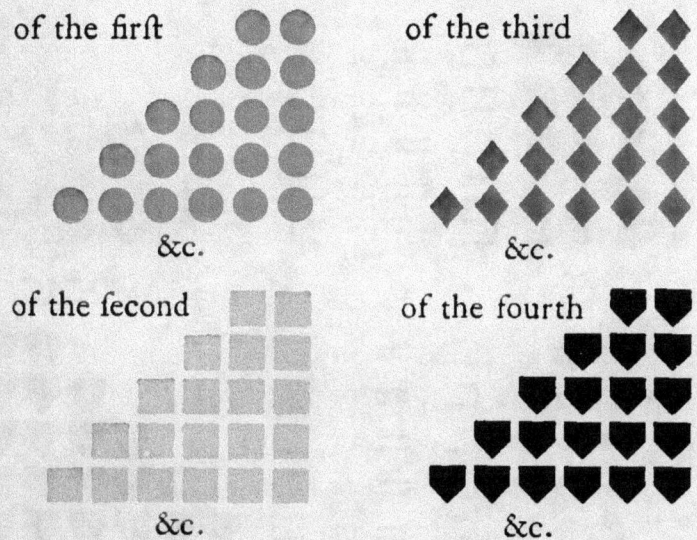

of the firſt

&c.

of the third

&c.

of the ſecond

&c.

of the fourth

&c.

Then taking every pair of equimultiples of the firſt and
third, and every pair of equimultiples of the ſecond and
fourth,

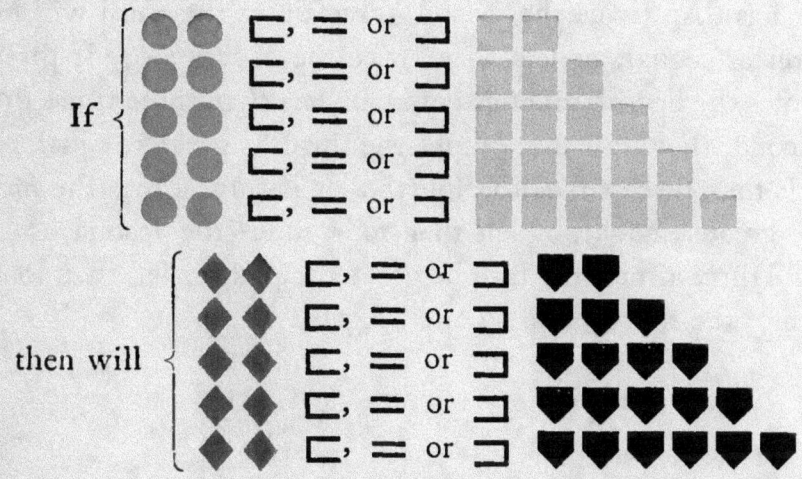

If

then will

That is, if twice the firſt be greater, equal, or leſs than twice the ſecond, twice the third will be greater, equal, or leſs than twice the fourth; or, if twice the firſt be greater, equal, or leſs than three times the ſecond, twice the third will be greater, equal, or leſs than three times the fourth, and ſo on, as above expreſſed.

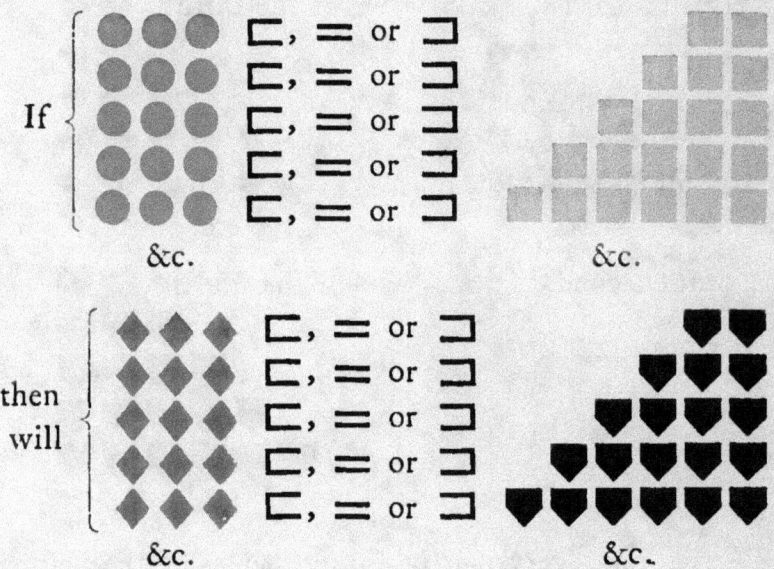

In other terms, if three times the firſt be greater, equal, or leſs than twice the ſecond, three times the third will be greater, equal, or leſs than twice the fourth; or, if three times the firſt be greater, equal, or leſs than three times the ſecond, then will three times the third be greater, equal, or leſs than three times the fourth; or if three times the firſt be greater, equal, or leſs than four times the ſecond, then will three times the third be greater, equal, or leſs than four times the fourth, and ſo on.   Again,

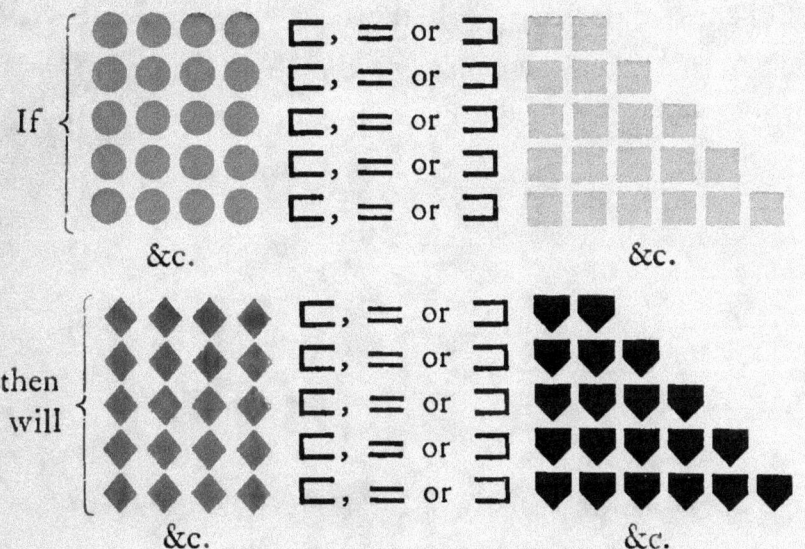

And so on, with any other equimultiples of the four magnitudes, taken in the fame manner.

Euclid expreffes this definition as follows :—

The firft of four magnitudes is faid to have the fame ratio to the fecond, which the third has to the fourth, when any equimultiples whatfoever of the firft and third being taken, and any equimultiples whatfoever of the fecond and fourth ; if the multiple of the firft be lefs than that of the fecond, the multiple of the third is alfo lefs than that of the fourth ; or, if the multiple of the firft be equal to that of the fecond, the multiple of the third is alfo equal to that of the fourth ; or, if the multiple of the firft be greater than that of the fecond, the multiple of the third is alfo greater than that of the fourth.

In future we fhall exprefs this definition generally, thus :

If M ● ⊏, = or ⊐ *m* ▢,

when M ◆ ⊏, = or ⊐ *m* ▼

x

Then we infer that ●, the firſt, has the ſame ratio to ■, the ſecond, which ◆, the third, has to ♥ the fourth: expreſſed in the ſucceeding demonſtrations thus:

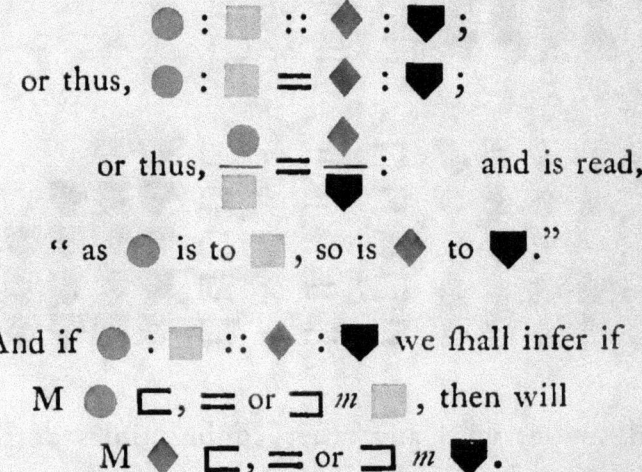

"as ● is to ■, so is ◆ to ♥."

And if ● : ■ :: ◆ : ♥ we ſhall infer if

M ● ⊏, = or ⊐ *m* ■, then will

M ◆ ⊏, = or ⊐ *m* ♥.

That is, if the firſt be to the ſecond, as the third is to the fourth; then if M times the firſt be greater than, equal to, or leſs than *m* times the ſecond, then ſhall M times the third be greater than, equal to, or leſs than *m* times the fourth, in which M and *m* are not to be conſidered particular multiples, but every pair of multiples whatever; nor are ſuch marks as ●, ♥, ■, &c. to be conſidered any more than repreſentatives of geometrical magnitudes.

The ſtudent ſhould thoroughly underſtand this definition before proceeding further.

F *the firſt of four magnitudes have the ſame ratio to the ſecond, which the third has to the fourth, then any equimultiples whatever of the firſt and third shall have the ſame ratio to any equimultiples of the ſecond and fourth; viz., the equimultiple of the firſt ſhall have the ſame ratio to that of the ſecond, which the equimultiple of the third has to that of the fourth.*

Let ● : ■ :: ◆ : ▼, then 3 ● : 2 ■ :: 3 ◆ : 2 ▼, every equimultiple of 3 ● and 3 ◆ are equimultiples of ● and ◆, and every equimultiple of 2 ■ and 2 ▼, are equimultiples of ■ and ▼ (B. 5, pr. 3.)

That is, M times 3 ● and M times 3 ◆ are equimultiples of ● and ◆, and *m* times 2 ■ and *m* 2 ▼ are equimultiples of 2 ■ and 2 ▼; but ● : ■ :: ◆ : ▼ (hyp); ∴ if M 3 ● ⊏, =, or ⊐ *m* 2 ■, then M 3 ◆ ⊏, =, or ⊐ *m* 2 ▼ (def. 5.) and therefore 3 ● : 2 ■ :: 3 ◆ : 2 ▼ (def. 5.)

The ſame reaſoning holds good if any other equimultiple of the firſt and third be taken, any other equimultiple of the ſecond and fourth.

∴ If the firſt four magnitudes, &c.

F one magnitude be the fame multiple of another, which a magnitude taken from the firſt is of a magnitude taken from the other, the remainder ſhall be the fame multiple of the remainder, that the whole is of the whole.

Let $\bigcirc\!\!\!\!\begin{array}{c}\bigcirc\\\bigcirc\end{array}\!\!\!\!\bigcirc = M' \begin{array}{c}\blacktriangle\\\blacksquare\end{array}$

and $\bigcup = M' \,\blacksquare,$

$\therefore\ \bigcirc\!\!\!\!\begin{array}{c}\bigcirc\\\bigcirc\end{array}\!\!\!\!\bigcirc$ minus $\bigcup = M'\begin{array}{c}\blacktriangle\\\blacksquare\end{array}$ minus $M'\,\blacksquare,$

$\therefore\ \bigcirc\!\!\!\!\bigcirc\!\!\!\!\bigcirc = M'\,(\begin{array}{c}\blacktriangle\\\blacksquare\end{array}$ minus $\blacksquare\,),$

and $\therefore\ \bigcirc\!\!\!\!\bigcirc\!\!\!\!\bigcirc = M'\,\blacktriangle.$

$\therefore$ If one magnitude, &c.

F *two magnitudes be equimultiples of two others,
and if equimultiples of these be taken from the first
two, the remainders are either equal to these others,
or equimultiples of them.*

Let ◇◇◇ = M′ ▪ ; and ◡◡ = M′ ▲ ;

then ◇◇ minus *m′* ▪ =

M′ ▪ minus *m′* ▪ = (M′ minus *m′*) ▪,

and ◡◡ minus *m′* ▲ = M′ ▲ minus *m′* ▲ =
(M′ minus *m′*) ▲ .

Hence, (M′ minus *m′*) ▪ and (M′ minus *m′*) ▲ are equimultiples of ▪ and ▲ , and equal to ▪ and ▲,
when M′ minus *m′* = 1.

∴ If two magnitudes be equimultiples, &c.

 F *the firſt of the four magnitudes has the ſame ratio to the ſecond which the third has to the fourth, then if the firſt be greater than the ſecond, the third is alſo greater than the fourth ; and if equal, equal ; if leſs, leſs.*

Let 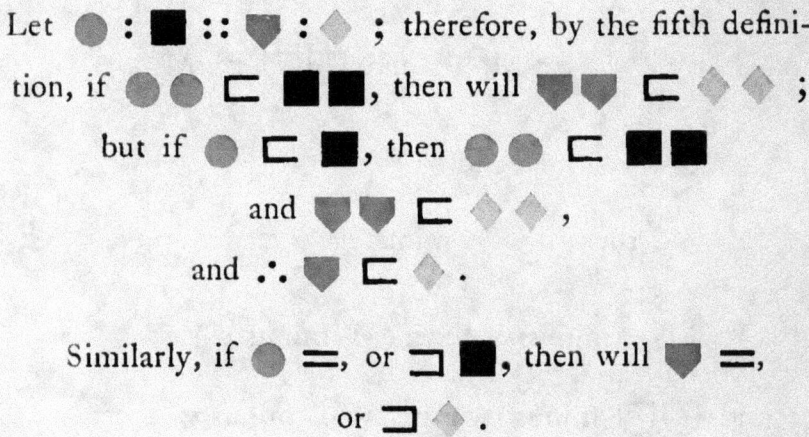 ; therefore, by the fifth definition, if ●● ⊏ ■■, then will ▼▼ ⊏ ◆◆ ;

but if ● ⊏ ■, then ●● ⊏ ■■

and ▼▼ ⊏ ◆◆ ,

and ∴ ▼ ⊏ ◆ .

Similarly, if ● =, or ⊐ ■, then will ▼ =,

or ⊐ ◆ .

∴ If the firſt of four, &c.

DEFINITION XIV.

GEOMETRICIANS make uſe of the technical term " Invertendo," by inverſion, when there are four proportionals, and it is inferred, that the ſecond is to the firſt as the fourth to the third.

Let A : B :: C : D, then, by " invertendo " it is inferred B : A :: D : C.

F *four magnitudes are proportionals, they are pro-portionals also when taken inversely.*

Let ■ : ◡ :: ■ : ◆ ,

then, inversely, ◡ : ■ :: ◆ : ■ .

If M ■ ⊐ *m* ◡, then M ■ ⊐ *m* ◆
by the fifth definition.

Let M ■ ⊐ *m* ◡, that is, *m* ◡ ⊏ M ■ .

∴ M ■ ⊐ *m* ◆ , or, *m* ◆ ⊏ M ■ ;

∴ if *m* ◡ ⊏ M ■ , then will *m* ◆ ⊏ M ■ .

In the same manner it may be shown,
that if *m* ◡ = or ⊐ M ■ ,
then will *m* ◆ =, or ⊐ M ■ ;
and therefore, by the fifth definition, we infer
that ◡ : ■ : ◆ : ■ .

∴ If four magnitudes, &c.

F *the firſt be the ſame multiple of the ſecond, or the ſame part of it, that the third is of the fourth; the firſt is to the ſecond, as the third is to the fourth.*

Let ▦ , the firſt, be the ſame multiple of ● , the ſecond,

that ◆◆ , the third, is of ▲ , the fourth.

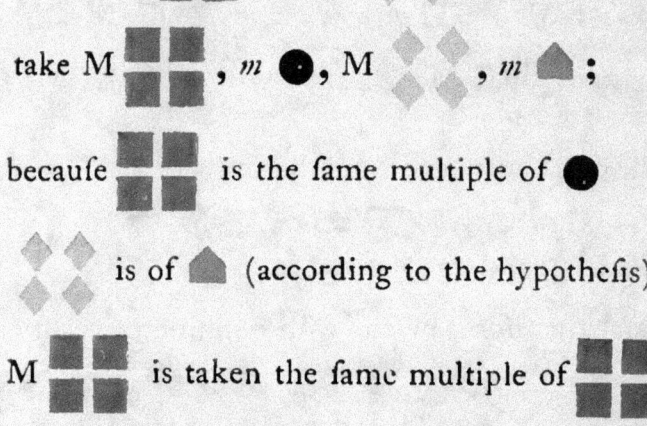

Then ▦ : ● :: ◆◆ : ▲

take M ▦ , *m* ● , M ◆◆ , *m* ▲ ;

becauſe ▦ is the ſame multiple of ●

that ◆◆ is of ▲ (according to the hypotheſis);

and M ▦ is taken the ſame multiple of ▦

that M ◆◆ is of ◆◆ ,

∴ (according to the third propoſition),

M ▦ is the ſame multiple of ●

that M ◆◆ is of ▲ .

Therefore, if M ⬛ be of ● a greater multiple than

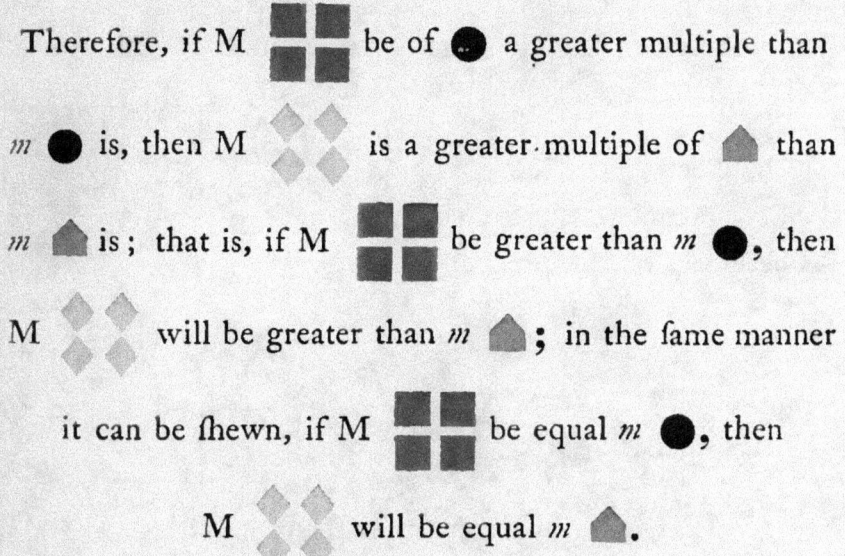

m ● is, then M ◆◆ is a greater multiple of ⬠ than

m ⬠ is; that is, if M ⬛ be greater than m ●, then

M ◆◆ will be greater than m ⬠; in the fame manner

it can be fhewn, if M ⬛ be equal m ●, then

M ◆◆ will be equal m ⬠.

And, generally, if M ⬛ ⊏, = or ⊐ m ●

then M ◆◆ will be ⊏, = or ⊐ m ⬠;

∴ by the fifth definition,

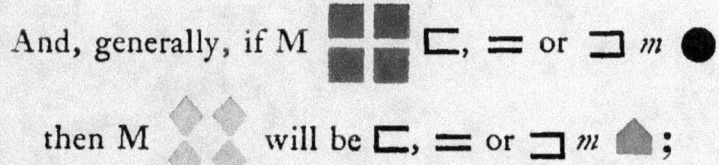

⬛ : ● :: ◆◆ : ⬠.

Next, let ● be the fame part of ⬛

that ⬠ is of ◆◆.

In this cafe alfo ● : ⬛ :: ⬠ : ◆◆.

For, becaufe

● is the fame part of ⬛ that ⬠ is of ◆◆,

Y

therefore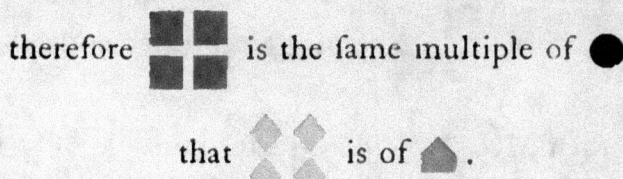

that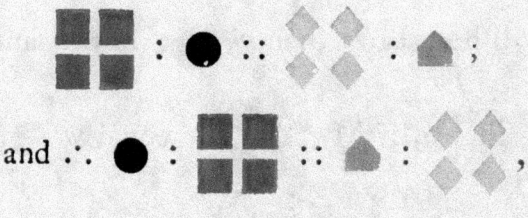

Therefore, by the preceding cafe,

and ∴

by propofition B.

∴ If the firft be the fame multiple, &c.

 F *the first be to the second as the third to the fourth, and if the first be a multiple, or a part of the second; the third is the same multiple, or the same part of the fourth.*

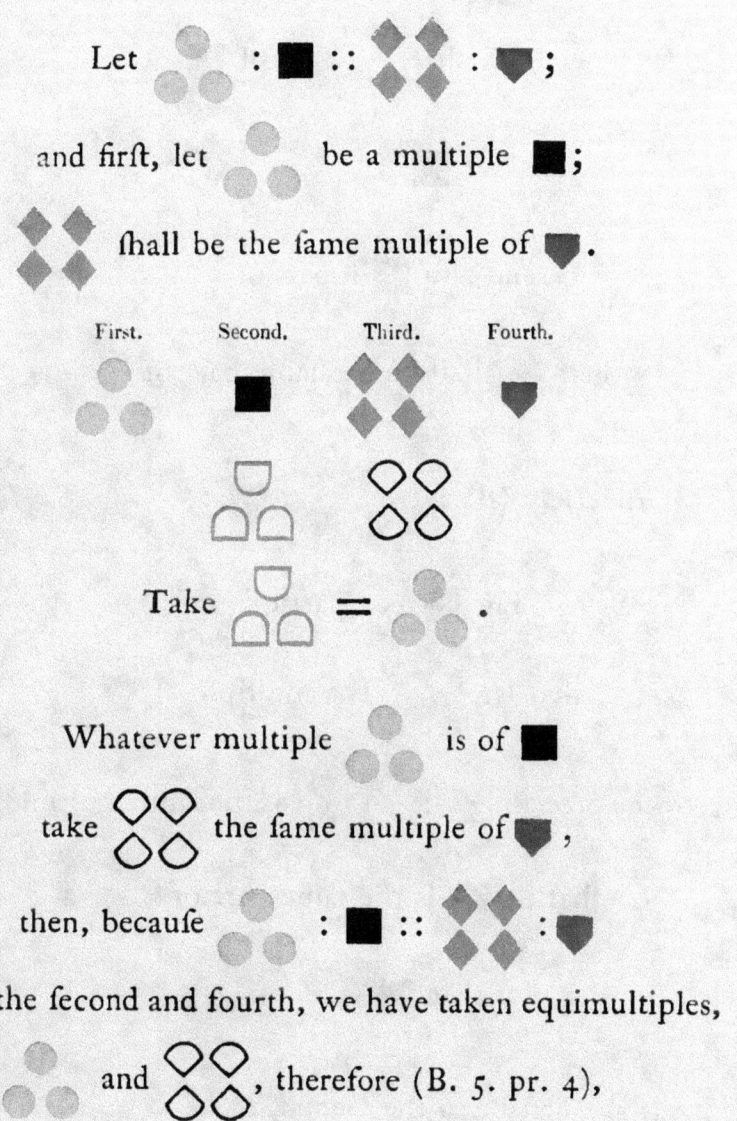

Let ⬤ : ■ :: ◆◆ : ▼;

and first, let ⬤ be a multiple ■;

◆◆ shall be the same multiple of ▼.

First. Second. Third. Fourth.

Take ᴅᴅᴅ = ⬤ .

Whatever multiple ⬤ is of ■

take ◯◯ the same multiple of ▼,

then, because ⬤ : ■ :: ◆◆ : ▼

and of the second and fourth, we have taken equimultiples,

⬤ and ◯◯ , therefore (B. 5. pr. 4),

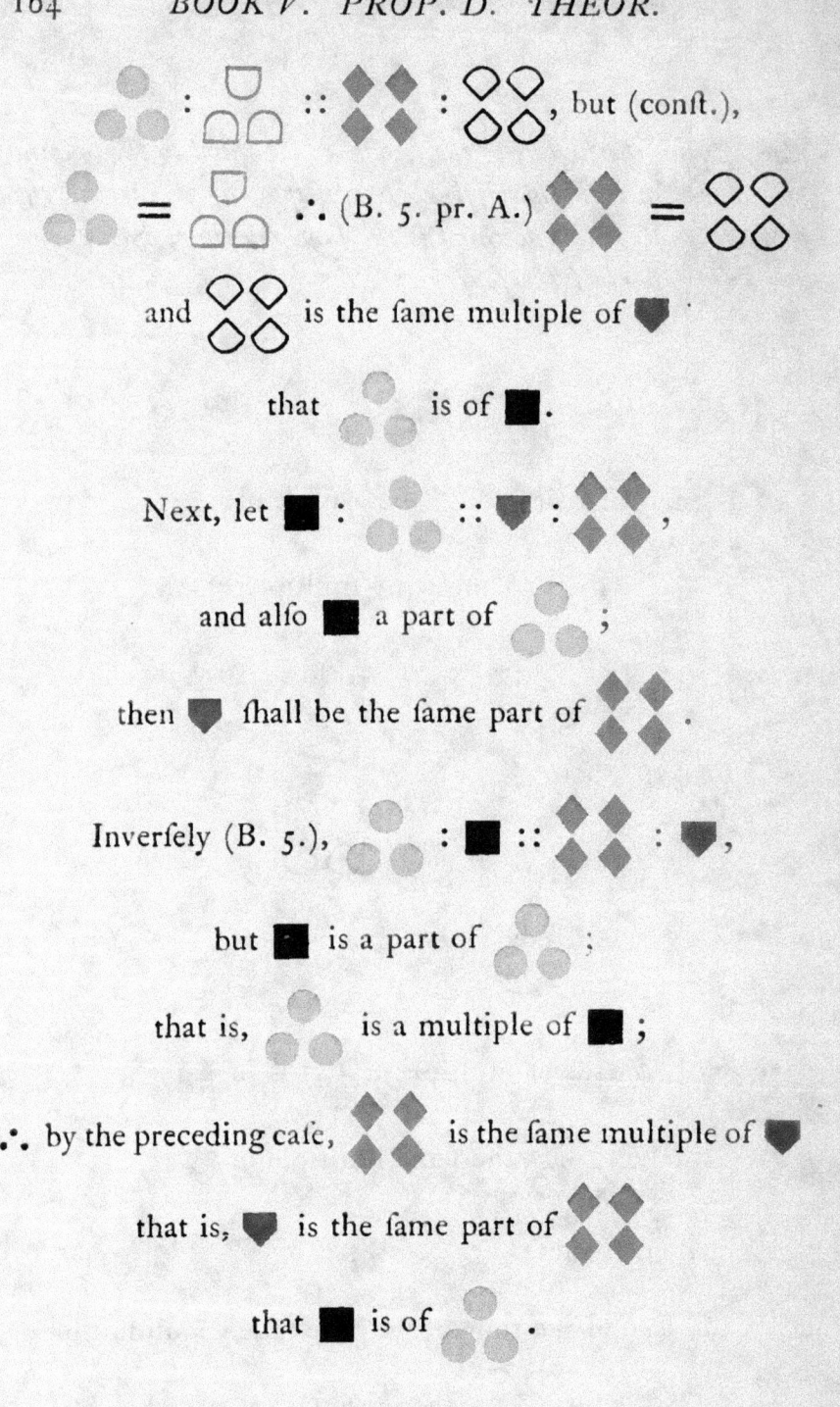

, but (conſt.),

∴ (B. 5. pr. A.)

and ⬭ is the ſame multiple of ▽

that ⬭ is of ■.

Next, let ■ : ⬭ :: ▽ : ◆◆ ,

and alſo ■ a part of ⬭ ;

then ▽ ſhall be the ſame part of ◆◆ .

Inverſely (B. 5.), ⬭ : ■ :: ◆◆ : ▽ ,

but ■ is a part of ⬭ ;

that is, ⬭ is a multiple of ■ ;

∴ by the preceding caſe, ◆◆ is the ſame multiple of ▽

that is, ▽ is the ſame part of ◆◆

that ■ is of ⬭ .

∴ If the firſt be to the ſecond, &c.

QUAL *magnitudes have the same ratio to the same magnitude, and the same has the same ratio to equal magnitudes.*

Let ● = ◆ and ▦ any other magnitude;
then ● : ▦ = ◆ : ▦ and ▦ : ● = ▦ : ◆.

Becaufe ● = ◆,

∴ M ● = M ◆ ;

∴ if M ● ⊏, = or ⊐ *m* ▦, then
M ◆ ⊏, = or ⊐ *m* ▦,
and ∴ ● : ▦ = ◆ : ▦ (B. 5. def. 5).

From the foregoing reafoning it is evident that,
if *m* ▦ ⊏, = or ⊐ M ●, then
m ▦ ⊏, = or ⊐ M ◆
∴ ▦ : ● = ▦ : ◆ (B. 5. def. 5).

∴ Equal magnitudes, &c.

DEFINITION VII.

WHEN of the equimultiples of four magnitudes (taken as in the fifth definition), the multiple of the firſt is greater than that of the ſecond, but the multiple of the third is not greater than the multiple of the fourth; then the firſt is ſaid to have to the ſecond a greater ratio than the third magnitude has to the fourth: and, on the contrary, the third is ſaid to have to the fourth a leſs ratio than the firſt has to the ſecond.

If, among the equimultiples of four magnitudes, compared as in the fifth definition, we ſhould find

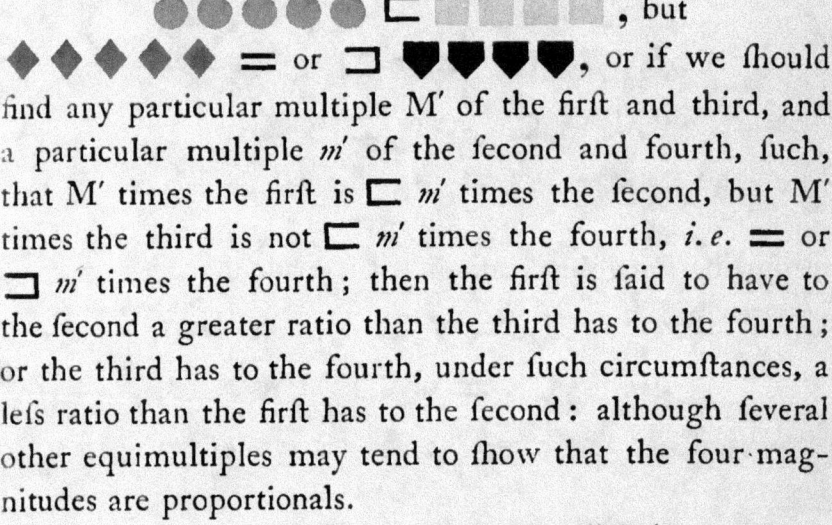

, but

, or if we ſhould find any particular multiple M′ of the firſt and third, and a particular multiple *m′* of the ſecond and fourth, ſuch, that M′ times the firſt is ⊏ *m′* times the ſecond, but M′ times the third is not ⊏ *m′* times the fourth, *i. e.* ═ or ⊐ *m′* times the fourth; then the firſt is ſaid to have to the ſecond a greater ratio than the third has to the fourth; or the third has to the fourth, under ſuch circumſtances, a leſs ratio than the firſt has to the ſecond: although ſeveral other equimultiples may tend to ſhow that the four-magnitudes are proportionals.

This definition will in future be expreſſed thus:—

If M′ ♥ ⊏ *m′* ∪, but M′ ■ ═ or ⊐ *m′* ◆,

then ♥ : ∪ ⊏ ■ : ◆ .

In the above general expreſſion, M′ and *m′* are to be conſidered particular multiples, not like the multiples M and *m* introduced in the fifth definition, which are in that definition confidered to be every pair of multiples that can be taken. It muft alfo be here obferved, that ◗, ◖, ◼, and the like fymbols are to be confidered merely the reprefentatives of geometrical magnitudes.

In a partial arithmetical way, this may be fet forth as follows:

Let us take the four numbers, 8, 7, 10, and 9.

First.	Second.	Third.	Fourth.
8	7	10	9
16	14	20	18
24	21	30	27
32	28	40	36
40	35	50	45
48	42	60	54
56	49	70	63
64	56	80	72
72	63	90	81
80	70	100	90
88	77	110	99
96	84	120	108
104	91	130	117
112	98	140	126
&c.	&c.	&c	&c.

Among the above multiples we find 16 ⊏ 14 and 20 ⊏ 18; that is, twice the firft is greater than twice the fecond, and twice the third is greater than twice the fourth; and 16 ⊐ 21 and 20 ⊐ 27; that is, twice the firft is lefs than three times the fecond, and twice the third is lefs than three times the fourth; and among the fame multiples we can find 72 ⊏ 56 and 90 ⊏ 72: that is, 9 times the firft is greater than 8 times the fecond, and 9 times the third is greater than 8 times the fourth. Many other equimul-

tiples might be selected, which would tend to fhow that
the numbers 8, 7, 10, 9, were proportionals, but they are
not, for we can find a multiple of the firſt ⊏ a multiple of
the ſecond, but the ſame multiple of the third that has been
taken of the firſt not ⊏ the ſame multiple of the fourth
which has been taken of the ſecond; for inſtance, 9 times
the firſt is ⊏ 10 times the ſecond, but 9 times the third is
not ⊏ 10 times the fourth, that is, 72 ⊏ 70, but 90
not ⊏ 90, or 8 times the firſt we find ⊏ 9 times the
ſecond, but 8 times the third is not greater than 9 times
the fourth, that is, 64 ⊏ 63, but 80 is not ⊏ 81. When
any ſuch multiples as theſe can be found, the firſt (8) is
ſaid to have to the ſecond (7) a greater ratio than the third
(10) has to the fourth (9), and on the contrary the third
(10) is ſaid to have to the fourth (9) a leſs ratio than the
firſt (8) has to the ſecond (7).

F *unequal magnitudes the greater has a greater ratio to the same than the less has : and the same magnitude has a greater ratio to the less than it has to the greater.*

Let ▲■ and ■ be two unequal magnitudes, and ● any other.

We shall first prove that ▲■ which is the greater of the two unequal magnitudes, has a greater ratio to ● than ■, the less, has to ● ;

that is, ▲■ : ● ⊏ ■ : ● ;

take M′ ▲■, *m′* ●, M′ ■, and *m′* ● ;

such, that M′ ▲ and M′ ■ shall be each ⊏ ● ;

also take *m′* ● the least multiple of ●,

which will make *m′* ● ⊏ M′ ■ = M′ ■ ;

∴ M′ ■ is not ⊏ *m′* ●,

but M′ ▲■ is ⊏ *m′* ●, for,

as *m′* ● is the first multiple which first becomes ⊏ M′ ■, than (*m′* minus 1) ● or *m′* ● minus ● is not ⊏ M′ ■, and ● is not ⊏ M′ ▲,

∴ *m′* ● minus ● + ● must be ⊐ M′ ■ + M′ ▲ ;

that is, *m′* ● must be ⊐ M′ ▲■ ;

∴ M′ ▲■ is ⊏ *m′* ● ; but it has been shown above that

z

169

M′ ■ is not ⊏ *m*′ ● , therefore, by the seventh definition,

▲ has to ● a greater ratio than ■ : ● .

Next we shall prove that ● has a greater ratio to ■ , the

less, than it has to ■ , the greater ;

or, ● : ■ ⊏ ● : ■ .

Take *m*′ ● , M′ ■ , *m*′ ● , and M′ ■ ,

the same as in the first case, such, that

M′ ▲ and M′ ■ will be each ⊏ ● , and *m*′ ● the least

multiple of ● , which first becomes greater

than M′ ■ = M′ ■ .

∴ *m*′ ● minus ● is not ⊏ M′ ■ ,

and ● is not ⊏ M′ ▲ ; consequently

m′ ● minus ● ✛ ● is ⊐ M′ ■ ✛ M′ ▲ ;

∴ *m*′ ● is ⊐ M′ ■ , and ∴ by the seventh definition,

● has to ■ a greater ratio than ● has to ■ .

∴ Of unequal magnitudes, &c.

The contrivance employed in this proposition for finding among the multiples taken, as in the fifth definition, a multiple of the first greater than the multiple of the second, but the same multiple of the third which has been taken of the first, not greater than the same multiple of the fourth which has been taken of the second, may be illustrated numerically as follows :—

The number 9 has a greater ratio to 7 than 8 has to 7 : that is, 9 : 7 ⊏ 8 : 7; or, 8 ✛ 1 : 7 ⊏ 8 : 7.

The multiple of 1, which firſt becomes greater than 7, is 8 times, therefore we may multiply the firſt and third by 8, 9, 10, or any other greater number; in this caſe, let us multiply the firſt and third by 8, and we have $64 + 8$ and 64: again, the firſt multiple of 7 which becomes greater than 64 is 10 times; then, by multiplying the ſecond and fourth by 10, we ſhall have 70 and 70; then, arranging theſe multiples, we have—

8 times the first.	10 times the second.	8 times the third.	10 times the fourth.
$64 + 8$	70	64	70

Conſequently $64 + 8$, or 72, is greater than 70, but 64 is not greater than 70, ∴ by the ſeventh definition, 9 has a greater ratio to 7 than 8 has to 7.

The above is merely illuſtrative of the foregoing demonſtration, for this property could be ſhown of theſe or other numbers very readily in the following manner; becauſe, if an antecedent contains its conſequent a greater number of times than another antecedent contains its conſequent, or when a fraction is formed of an antecedent for the numerator, and its conſequent for the denominator be greater than another fraction which is formed of another antecedent for the numerator and its conſequent for the denominator, the ratio of the firſt antecedent to its conſequent is greater than the ratio of the laſt antecedent to its conſequent.

Thus, the number 9 has a greater ratio to 7, than 8 has to 7, for $\frac{9}{7}$ is greater than $\frac{8}{7}$.

Again, $17 : 19$ is a greater ratio than $13 : 15$, becauſe $\frac{17}{19} = \frac{17 \times 15}{19 \times 15} = \frac{255}{285}$, and $\frac{13}{15} = \frac{13 \times 19}{15 \times 19} = \frac{247}{285}$, hence it is evident that $\frac{255}{285}$ is greater than $\frac{247}{285}$, ∴ $\frac{17}{19}$ is greater than

$\frac{13}{15}$, and, according to what has been above shown, 17 has to 19 a greater ratio than 13 has to 15.

So that the general terms upon which a greater, equal, or less ratio exists are as follows :—

If $\frac{A}{B}$ be greater than $\frac{C}{D}$, A is said to have to B a greater ratio than C has to D ; if $\frac{A}{B}$ be equal to $\frac{C}{D}$, then A has to B the same ratio which C has to D ; and if $\frac{A}{B}$ be less than $\frac{C}{D}$, A is said to have to B a less ratio than C has to D.

The student should understand all up to this proposition perfectly before proceeding further, in order fully to comprehend the following propositions of this book. We therefore strongly recommend the learner to commence again, and read up to this slowly, and carefully reason at each step, as he proceeds, particularly guarding against the mischievous system of depending wholly on the memory. By following these instructions, he will find that the parts which usually present considerable difficulties will present no difficulties whatever, in prosecuting the study of this important book.

AGNITUDES *which have the same ratio to the same magnitude are equal to one another; and those to which the same magnitude has the same ratio are equal to one another.*

Let ◆ : ▨ :: ⬤ : ▨, then ◆ = ⬤.

For, if not, let ◆ ⊏ ⬤, then will

◆ : ▨ ⊏ ⬤ : ▨ (B. 5. pr. 8),

which is absurd according to the hypothesis.

∴ ◆ is not ⊏ ⬤.

In the same manner it may be shown, that

⬤ is not ⊏ ◆,

∴ ◆ = ⬤.

Again, let ▨ : ◆ :: ▨ : ⬤, then will ◆ = ⬤.

For (invert.) ◆ : ▨ :: ⬤ : ▨,

therefore, by the first case, ◆ = ⬤.

∴ Magnitudes which have the same ratio, &c.

This may be shown otherwise, as follows :—

Let $A : B = A : C$, then $B = C$, for, as the fraction $\frac{A}{B} =$ the fraction $\frac{A}{C}$, and the numerator of one equal to the numerator of the other, therefore the denominator of these fractions are equal, that is $B = C$.

Again, if $B : A = C : A$, $B = C$. For, as $\frac{B}{A} = \frac{C}{A}$, B must $= C$.

HAT *magnitude which has a greater ratio than another has unto the same magnitude, is the greater of the two : and that magnitude to which the same has a greater ratio than it has unto another magnitude, is the less of the two.*

Let ▼ : ■ ⊏ ● : ■ , then ▼ ⊏ ● .

For if not, let ▼ = or ⊐ ● ;

then, ▼ : ■ = ● : ■ (B. 5. pr. 7) or

▼ : ■ ⊐ ● : ■ (B. 5. pr. 8) and (invert.),
which is abſurd according to the hypotheſis.

∴ ▼ is not = or ⊐ ● ; and

∴ ▼ muſt be ⊏ ● .

Again, let ■ : ● ⊏ ■ : ▼ ,

then, ● ⊐ ▼ .

For if not, ● muſt be ⊏ or = ▼ ,

then ■ : ● ⊐ ■ : ▼ (B. 5. pr. 8) and (invert.) ;

or ■ : ● = ■ : ▼ (B. 5. pr. 7), which is abſurd (hyp.) ;

∴ ● is not ⊏ or = ▼ ,

and ∴ ● muſt be ⊐ ▼ .

∴ That magnitude which has, &c.

ATIOS *that are the same to the same ratio, are the same to each other.*

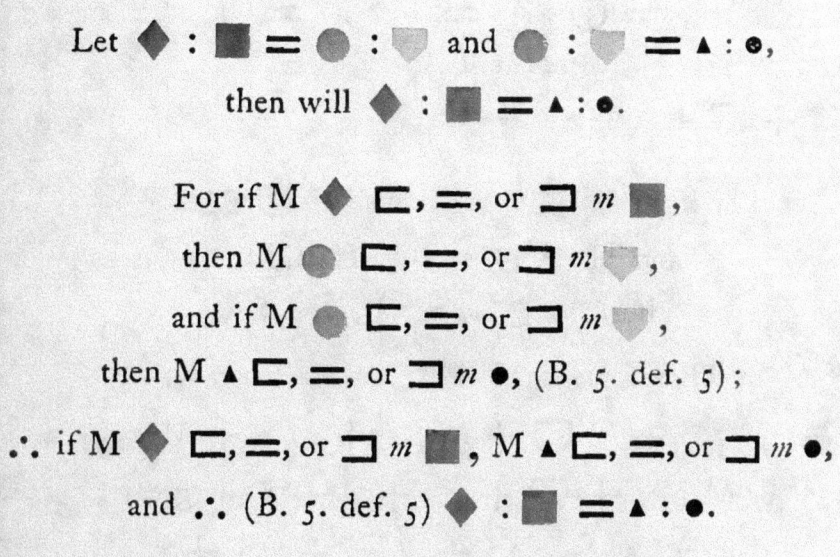

∴ Ratios that are the fame, &c.

F *any number of magnitudes be proportionals, as one of the antecedents is to its confequent, fo fhall all the antecedents taken together be to all the confequents.*

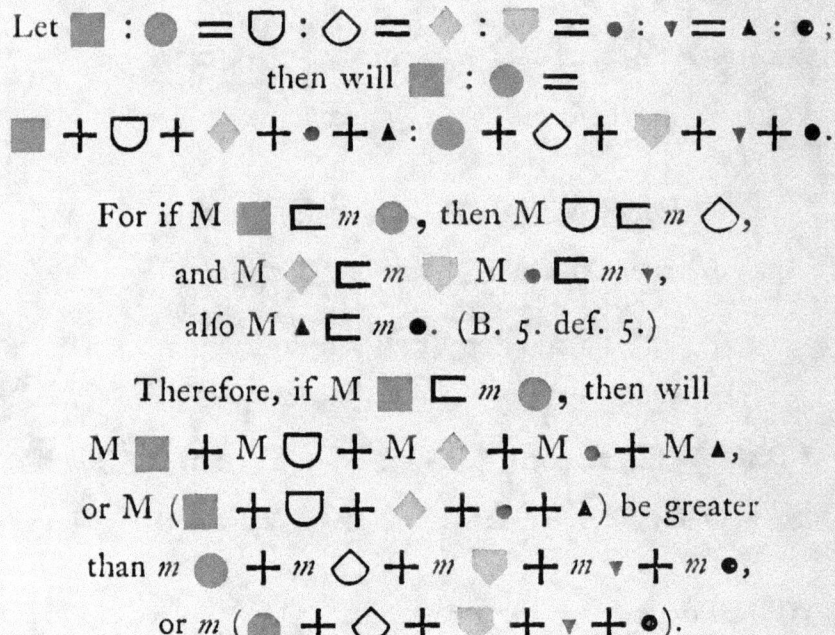

In the fame way it may be fhown, if M times one of the antecedents be equal to or lefs than *m* times one of the confequents, M times all the antecedents taken together, will be equal to or lefs than *m* times all the confequents taken together. Therefore, by the fifth definition, as one of the antecedents is to its confequent, fo are all the antecedents taken together to all the confequents taken together.

∴ If any number of magnitudes, &c.

F *the firſt has to the ſecond the ſame ratio which the third has to the fourth, but the third to the fourth a greater ratio than the fifth has to the ſixth ; the firſt ſhall alſo have to the ſecond a greater ratio than the fifth to the ſixth.*

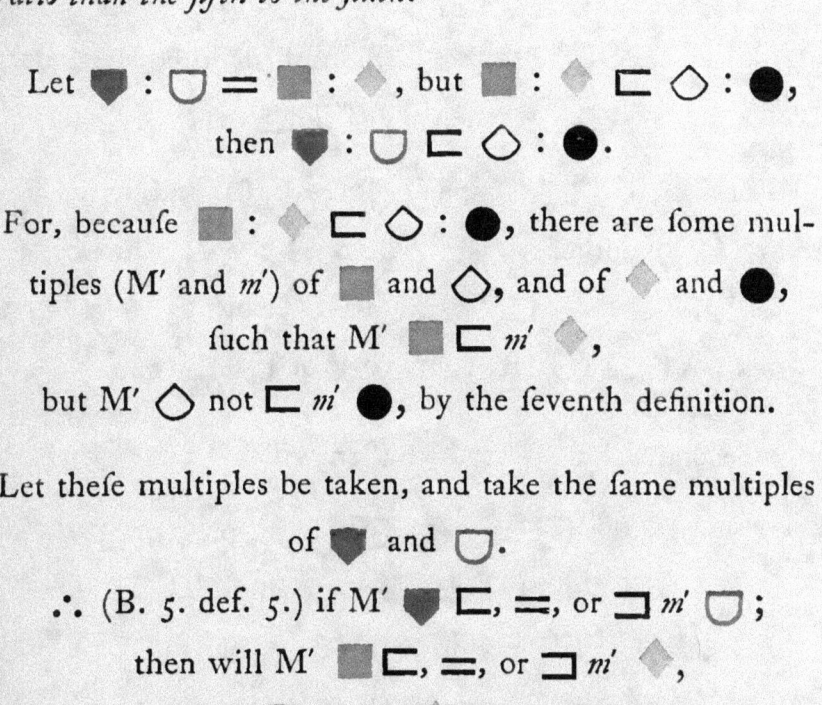

Let ⬟ : ⌒ = ⬛ : ◆, but ⬛ : ◆ ⊏ ◇ : ●,

then ⬟ : ⌒ ⊏ ◇ : ●.

For, becauſe ⬛ : ◆ ⊏ ◇ : ●, there are ſome mul-

tiples (M′ and *m′*) of ⬛ and ◇, and of ◆ and ●,

ſuch that M′ ⬛ ⊏ *m′* ◆,

but M′ ◇ not ⊏ *m′* ●, by the ſeventh definition.

Let theſe multiples be taken, and take the ſame multiples

of ⬟ and ⌒.

∴ (B. 5. def. 5.) if M′ ⬟ ⊏, =, or ⊐ *m′* ⌒ ;

then will M′ ⬛ ⊏, =, or ⊐ *m′* ◆,

but M′ ⬛ ⊏ *m′* ◆ (conſtruction) ;

∴ M′ ⬟ ⊏ *m′* ⌒,

but M′ ◇ is not ⊏ *m′* ● (conſtruction) ;

and therefore by the ſeventh definition,

⬟ : ⌒ ⊏ ◇ : ●.

∴ If the firſt has to the ſecond, &c.

A A

F *the firſt has the ſame ratio to the ſecond which the third has to the fourth; then, if the firſt be greater than the third, the ſecond ſhall be greater than the fourth; and if equal, equal; and if leſs, leſs.*

Let ▼ : ◡ :: ■ : ◆, and firſt ſuppoſe ▼ ⊏ ■, then will ◡ ⊏ ◆.

For ▼ : ◡ ⊏ ■ : ◡ (B. 5. pr. 8), and by the hypotheſis, ▼ : ◡ = ■ : ◆ ;

∴ ■ : ◆ ⊏ ■ : ◡ (B. 5. pr. 13),

∴ ◆ ⊐ ◡ (B. 5. pr. 10.), or ◡ ⊏ ◆.

Secondly, let ▼ = ■, then will ◡ = ◆.

For ▼ : ◡ = ■ : ◡ (B. 5. pr. 7),

and ▼ : ◡ = ■ : ◆ (hyp.) ;

∴ ■ : ◡ = ■ : ◆ (B. 5. pr. 11),

and ∴ ◡ = ◆ (B. 5, pr. 9).

Thirdly, if ▼ ⊐ ■, then will ◡ ⊐ ◆ ;

becauſe ■ ⊏ ▼ and ■ : ◆ = ▼ : ◡ ;

∴ ◆ ⊏ ◡, by the firſt caſe,

that is, ◡ ⊐ ◆.

∴ If the firſt has the ſame ratio, &c.

AGNITUDES *have the same ratio to one another which their equimultiples have.*

Let ● and ■ be two magnitudes;

then, ● : ■ :: M′ ● : M′ ■.

For ● : ■ = ● : ■

= ● : ■

= ● : ■

∴ ● : ■ :: 4 ● : 4 ■. (B. 5. pr. 12).

And as the same reasoning is generally applicable, we have

● : ■ :: M′ ● : M′ ■.

∴ Magnitudes have the same ratio, &c.

DEFINITION XIII.

THE technical term permutando, or alternando, by permutation or alternately, is ufed when there are four proportionals, and it is inferred that the firft has the fame ratio to the third which the fecond has to the fourth; or that the firft is to the third as the fecond is to the fourth: as is fhown in the following propofition:—

Let ⬡ : ◆ :: ▼ : ■,

by " permutando" or " alternando" it is

inferred ⬡ : ▼ :: ◆ : ■.

It may be neceffary here to remark that the magnitudes ⬡, ◆, ▼, ■, muft be homogeneous, that is, of the fame nature or fimilitude of kind; we muft therefore, in fuch cafes, compare lines with lines, furfaces with furfaces, folids with folids, &c. Hence the ftudent will readily perceive that a line and a furface, a furface and a folid, or other heterogenous magnitudes, can never ftand in the relation of antecedent and confequent.

F *four magnitudes of the same kind be proportionals, they are also proportionals when taken alternately.*

Let ▼ : ▽ :: ▦ : ◆, then ▼ : ▦ :: ▽ : ◆.

For M ▼ : M ▽ :: ▼ : ▽ (B. 5. pr. 15),

and M ▼ : M ▽ :: ▦ : ◆ (hyp.) and (B. 5. pr. 11);

also *m* ▦ : *m* ◆ :: ▦ : ◆ (B. 5. pr. 15);

∴ M ▼ : M ▽ :: *m* ▦ : *m* ◆ (B. 5. pr. 14),

and ∴ if M ▼ ⊏, =, or ⊐ *m* ▦,

then will M ▽ ⊏, =, or ⊐ *m* ◆ (B. 5. pr. 14);

therefore, by the fifth definition,

▼ : ▦ :: ▽ : ◆.

∴ If four magnitudes of the same kind, &c.

DEFINITION XVI.

DIVIDENDO, by divifion, when there are four proportionals, and it is inferred, that the excefs of the firft above the fecond is to the fecond, as the excefs of the third above the fourth, is to the fourth.

Let A : B :: C : D ;

by "dividendo" it is inferred

A minus B : B :: C minus D : D.

According to the above, A is fuppofed to be greater than B, and C greater than D ; if this be not the cafe, but to have B greater than A, and D greater than C, B and D can be made to ftand as antecedents, and A and C as confequents, by "invertion"

B : A :: D : C ;

then, by "dividendo," we infer

B minus A : A :: D minus C : C.

IF magnitudes, taken jointly, be proportionals, they *shall also be proportionals when taken separately: that is, if two magnitudes together have to one of them the same ratio which two others have to one of these, the remaining one of the first two shall have to the other the same ratio which the remaining one of the last two has to the other of these.*

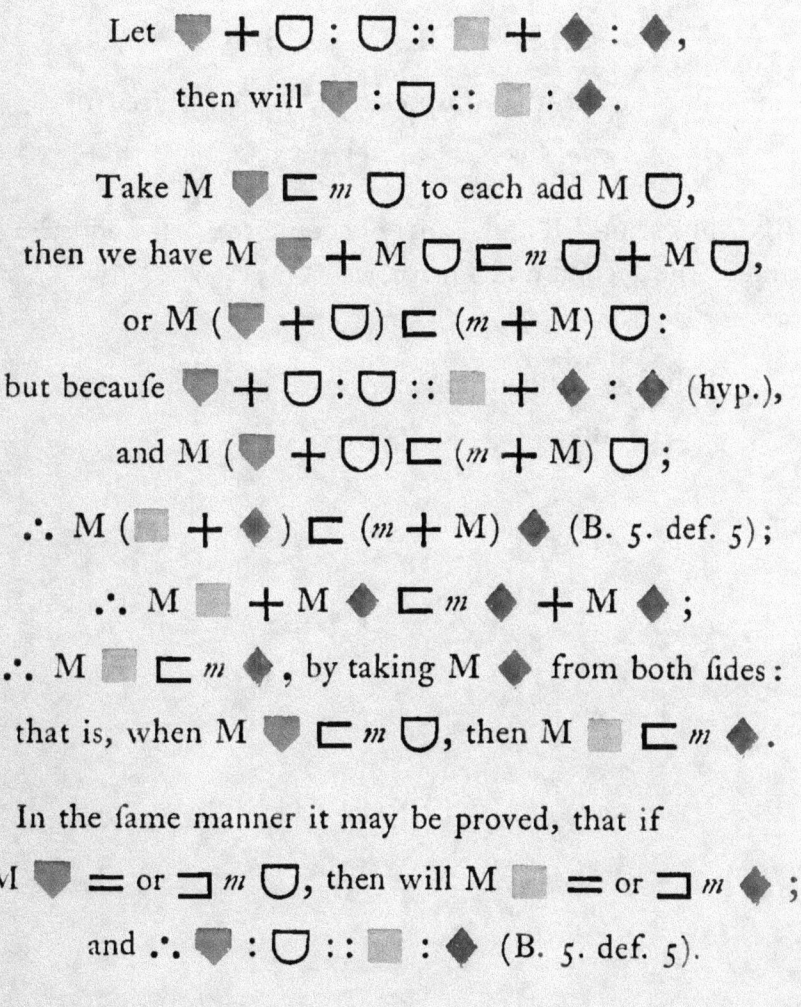

Let ▼ + ◡ : ◡ :: ■ + ◆ : ◆,

then will ▼ : ◡ :: ■ : ◆.

Take M ▼ ⊏ *m* ◡ to each add M ◡,

then we have M ▼ + M ◡ ⊏ *m* ◡ + M ◡,

or M (▼ + ◡) ⊏ (*m* + M) ◡ :

but because ▼ + ◡ : ◡ :: ■ + ◆ : ◆ (hyp.),

and M (▼ + ◡) ⊏ (*m* + M) ◡ ;

∴ M (■ + ◆) ⊏ (*m* + M) ◆ (B. 5. def. 5);

∴ M ■ + M ◆ ⊏ *m* ◆ + M ◆ ;

∴ M ■ ⊏ *m* ◆ , by taking M ◆ from both sides:

that is, when M ▼ ⊏ *m* ◡ , then M ■ ⊏ *m* ◆ .

In the same manner it may be proved, that if

M ▼ = or ⊐ *m* ◡ , then will M ■ = or ⊐ *m* ◆ ;

and ∴ ▼ : ◡ :: ■ : ◆ (B. 5. def. 5).

∴ If magnitudes taken jointly, &c.

DEFINITION XV.

THE term componendo, by compofition, is ufed when there are four proportionals; and it is inferred that the firſt together with the ſecond is to the ſecond as the third together with the fourth is to the fourth.

Let A : B :: C : D ;

then, by the term " componendo," it is inferred that

A + B : B :: C + D : D.

By " invertion" B and D may become the firſt and third, A and C the ſecond and fourth, as

B : A :: D : C,

then, by " componendo," we infer that

B + A : A :: D + C : C.

F *magnitudes, taken separately, be proportionals, they shall also be proportionals when taken jointly: that is, if the first be to the second as the third is to the fourth, the first and second together shall be to the second as the third and fourth together is to the fourth.*

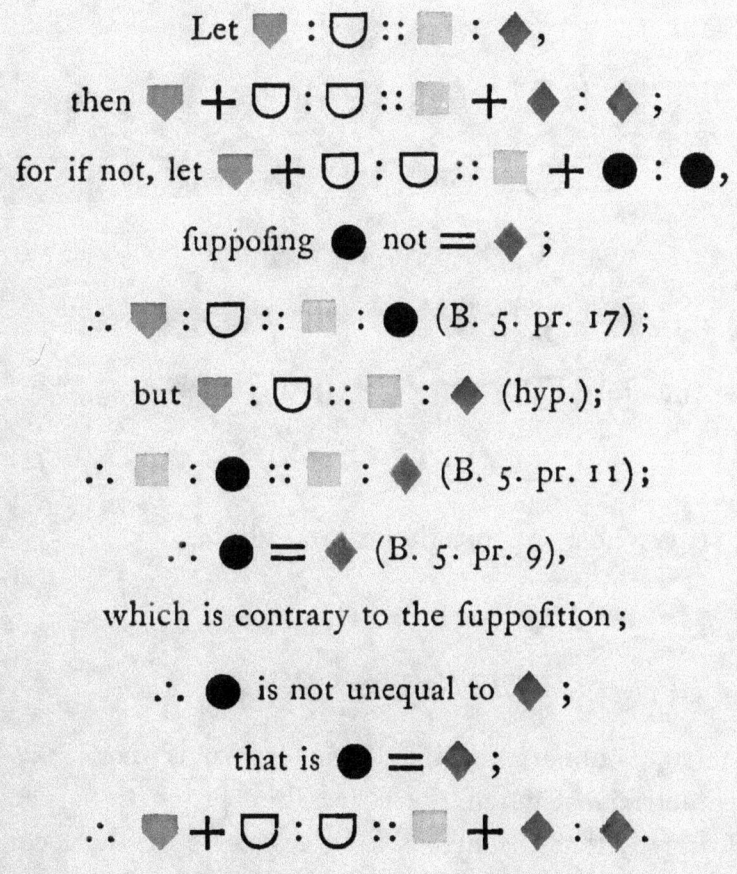

which is contrary to the supposition;

∴ ● is not unequal to ◆ ;

that is ● = ◆ ;

∴ ▼ + ⛝ : ⛝ :: ▦ + ◆ : ◆.

∴ If magnitudes, taken separately, &c.

 F *a whole magnitude be to a whole, as a magnitude taken from the firſt, is to a magnitude taken from the other ; the remainder ſhall be to the remainder, as the whole to the whole.*

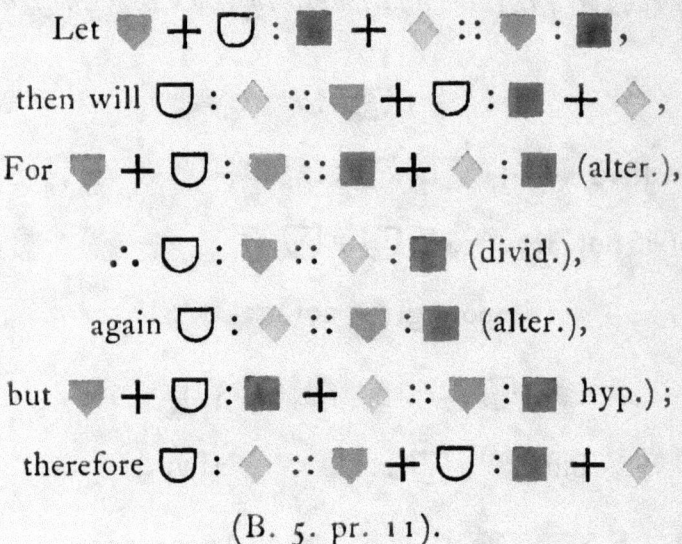

Let ▼ + ◡ : ■ + ◆ :: ▼ : ■ ,

then will ◡ : ◆ :: ▼ + ◡ : ■ + ◆ ,

For ▼ + ◡ : ▼ :: ■ + ◆ : ■ (alter.),

∴ ◡ : ▼ :: ◆ : ■ (divid.),

again ◡ : ◆ :: ▼ : ■ (alter.),

but ▼ + ◡ : ■ + ◆ :: ▼ : ■ hyp.) ;

therefore ◡ : ◆ :: ▼ + ◡ : ■ + ◆

(B. 5. pr. 11).

∴ If a whole magnitude be to a whole, &c.

DEFINITION XVII.

T HE term "convertendo," by converſion, is made uſe of by geometricians, when there are four proportionals, and it is inferred, that the firſt is to its exceſs above the ſecond, as the third is to its exceſs above the fourth. See the following propoſition :—

F *four magnitudes be proportionals, they are also proportionals by conversion: that is, the first is to its excess above the second, as the third to its excess above the fourth.*

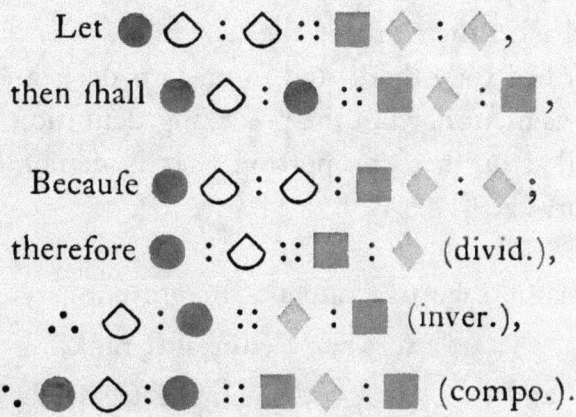

∴ If four magnitudes, &c.

DEFINITION XVIII.

" Ex æquali" (fc. diſtantiâ), or ex æquo, from equality of diſtance : when there is any number of magnitudes more than two, and as many others, ſuch that they are proportionals when taken two and two of each rank, and it is inferred that the firſt is to the laſt of the firſt rank of magnitudes, as the firſt is to the laſt of the others : " of this there are the two following kinds, which ariſe from the different order in which the magnitudes are taken, two and two."

DEFINITION XIX.

" Ex æquali," from equality. This term is uſed ſimply by itſelf, when the firſt magnitude is to the ſecond of the firſt rank, as the firſt to the ſecond of the other rank ; and as the ſecond is to the third of the firſt rank, ſo is the ſecond to the third of the other ; and ſo on in order : and the inference is as mentioned in the preceding definition ; whence this is called ordinate proportion. It is demonſtrated in Book 5. pr. 22.

Thus, if there be two ranks of magnitudes,

A, B, C, D, E, F, the firſt rank,

and L, M, N, O, P, Q, the ſecond,

ſuch that A : B :: L : M, B : C :: M : N,

C : D :: N : O, D : E :: O : P, E : F :: P : Q;

we infer by the term " ex æquali " that

A : F :: L : Q.

DEFINITION XX.

" Ex æquali in proportione perturbatâ feu inordinatâ," from equality in perturbate, or diforderly proportion. This term is ufed when the firft magnitude is to the fecond of the firft rank as the laft but one is to the laft of the fecond rank ; and as the fecond is to the third of the firft rank, fo is the laft but two to the laft but one of the fecond rank ; and as the third is to the fourth of the firft rank, fo is the third from the laft to the laft but two of the fecond rank ; and fo on in a crofs order : and the inference is in the 18th definition.   It is demonftrated in B. 5. pr. 23.

Thus, if there be two ranks of magnitudes,

A, B, C, D, E, F, the firft rank,

and L, M, N, O, P, Q, the fecond,

fuch that A : B :: P : Q, B : C :: O : P,

C : D :: N : O, D : E :: M : N, E : F :: L : M ;

the term " ex æquali in proportione perturbatâ feu inordi-

natâ" infers that

A : F :: L : Q.

F *there be three magnitudes, and other three, which, taken two and two, have the same ratio ; then, if the first be greater than the third, the fourth shall be greater than the sixth ; and if equal, equal ; and if less, less.*

Let ▼, ◡, ▦, be the first three magnitudes,

and ◆, ◇, ●, be the other three,

such that ▼ : ◡ :: ◆ : ◇, and ◡ : ▦ :: ◇ : ●.

Then, if ▼ ⊏, =, or ⊐ ▦, then will ◆ ⊏, =,

or ⊐ ●.

From the hypothesis, by alternando, we have

▼ : ◆ :: ◡ : ◇,

and ◡ : ◇ :: ▦ : ● ;

∴ ▼ : ◆ :: ▦ : ● (B. 5. pr. 11);

∴ if ▼ ⊏, =, or ⊐ ▦, then will ◆ ⊏, =,

or ⊐ ● (B. 5. pr. 14).

∴ If there be three magnitudes, &c.

F *there be three magnitudes, and other three which have the fame ratio, taken two and two, but in a crofs order ; then if the firft magnitude be greater than the third, the fourth fhall be greater than the fixth ; and if equal, equal ; and if lefs, lefs.*

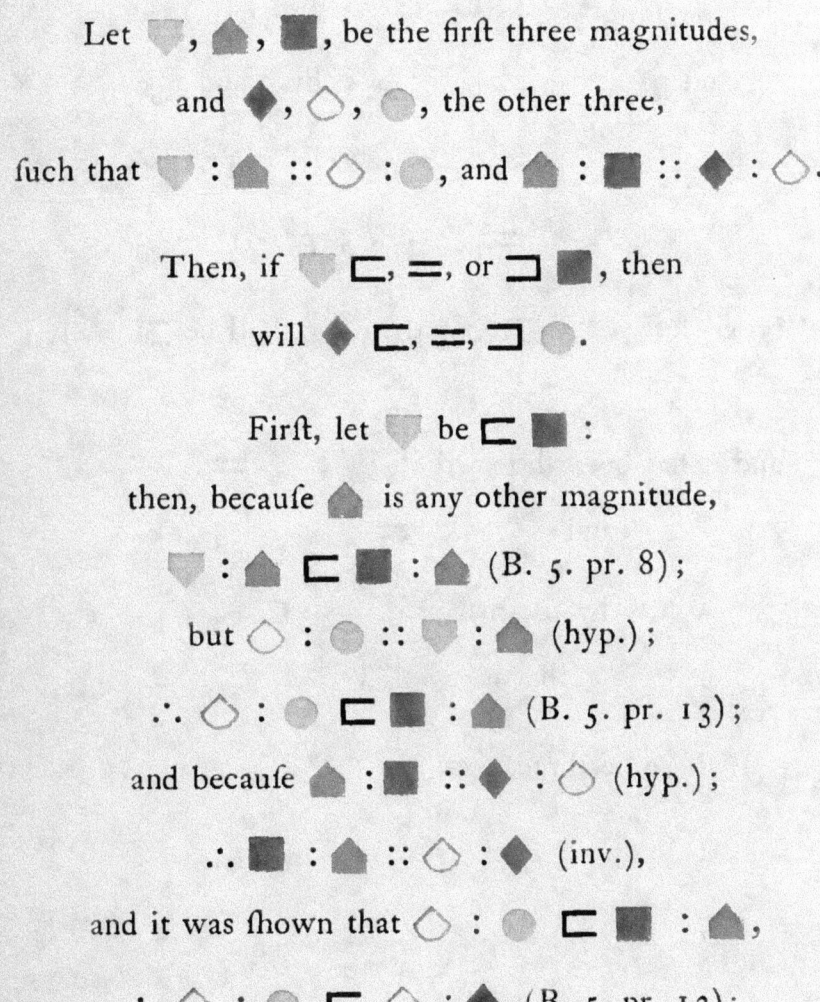

Let ▼, ⬠, ■, be the firft three magnitudes,

and ◆, △, ●, the other three,

fuch that ▼ : ⬠ :: △ : ●, and ⬠ : ■ :: ◆ : △.

Then, if ▼ ⊏, =, or ⊐ ■, then

will ◆ ⊏, =, ⊐ ●.

First, let ▼ be ⊏ ■ :

then, becaufe ⬠ is any other magnitude,

▼ : ⬠ ⊏ ■ : ⬠ (B. 5. pr. 8) ;

but △ : ● :: ▼ : ⬠ (hyp.) ;

∴ △ : ● ⊏ ■ : ⬠ (B. 5. pr. 13) ;

and becaufe ⬠ : ■ :: ◆ : △ (hyp.) ;

∴ ■ : ⬠ :: △ : ◆ (inv.),

and it was fhown that △ : ● ⊏ ■ : ⬠,

∴ △ : ● ⊏ △ : ◆ (B. 5. pr. 13) ;

∴ ● ⊐ ◆,

that is ◆ ⊏ ●.

Secondly, let ▼ = ■; then fhall ◆ = ●.

For becaufe ▼ = ■,

▼ : ▲ = ■ : ▲ (B. 5. pr. 7);

but ▼ : ▲ = ◇ : ● (hyp.),

and ■ : ▲ = ◇ : ◆ (hyp. and inv.),

∴ ◇ : ● = ◇ : ◆ (B. 5. pr. 11),

∴ ◆ = ● (B. 5. pr. 9).

Next, let ▼ be ⊐ ■, then ◆ fhall be ⊐ ● ;

for ■ ⊏ ▼,

and it has been fhown that ■ : ▲ = ◇ : ◆,

and ▲ : ▼ = ● : ◇ ;

∴ by the firft cafe ● is ⊏ ◆,

that is, ◆ ⊐ ●.

∴ If there be three, &c.

F *there be any number of magnitudes, and as many others, which, taken two and two in order, have the same ratio; the first shall have to the last of the first magnitudes the same ratio which the first of the others has to the last of the same.*

N.B.—*This is usually cited by the words "ex æquali," or "ex æquo."*

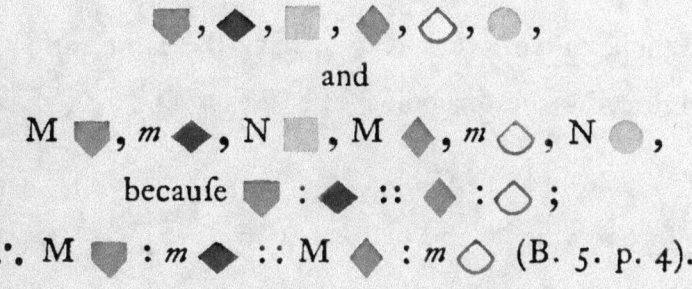

For the same reason

and because there are three magnitudes,

c c

M ▼, *m* ◆, N ■,

and other three, M ◆, *m* ◇, N ●,

which, taken two and two, have the fame ratio;

∴ if M ▼ ⊏, =, or ⊐ N ■

then will M ◆ ⊏, =, or ⊐ N ●, by (B. 5. pr. 20);

and ∴ ▼ : ■ :: ◆ : ● (def. 5).

Next, let there be four magnitudes, ▼, ◆, ■, ◆,

and other four, ◇, ●, ▬, ▲,

which, taken two and two, have the fame ratio,

that is to fay, ▼ : ◆ :: ◇ : ●,

◆ : ■ :: ● : ▬,

and ■ : ◆ :: ▬ : ▲,

then fhall ▼ : ◆ :: ◇ : ▲;

for, becaufe ▼, ◆, ■, are three magnitudes,

and ◇, ●, ▬, other three,

which, taken two and two, have the fame ratio;

therefore, by the foregoing cafe, ▼ : ■ :: ◇ : ▬,

but ■ : ◆ :: ▬ : ▲;

therefore again, by the firft cafe, ▼ : ◆ :: ◇ : ▲;

and fo on, whatever the number of magnitudes be.

∴ If there be any number, &c.

F *there be any number of magnitudes, and as many others, which, taken two and two in a cross order, have the same ratio ; the first shall have to the last of the first magnitudes the same ratio which the first of the others has to the last of the same.*

N.B.—*This is usually cited by the words " ex æquali in proportione perturbatâ ;" or " ex æquo perturbato."*

First, let there be three magnitudes, ▽ , ◖ , ■ ,

and other three, ◆ , △ , ● ,

which, taken two and two in a cross order,

have the same ratio ;

that is, ▽ : ◖ :: △ : ● ,

and ◖ : ■ :: ◆ : △ ,

then shall ▽ : ■ :: ◆ : ● .

Let these magnitudes and their respective equimultiples be arranged as follows :—

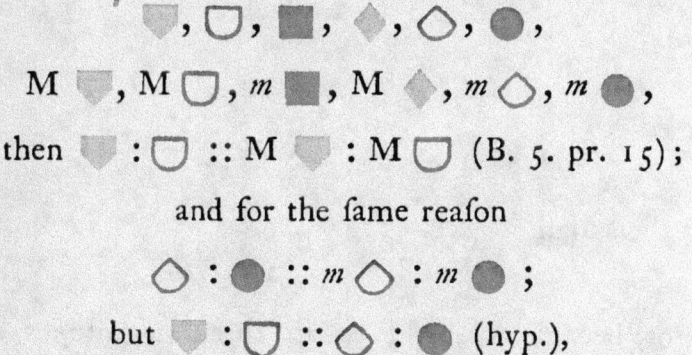

▽ , ◖ , ■ , ◆ , △ , ● ,

M ▽ , M ◖ , *m* ■ , M ◆ , *m* △ , *m* ● ,

then ▽ : ◖ :: M ▽ : M ◖ (B. 5. pr. 15) ;

and for the same reason

△ : ● :: *m* △ : *m* ● ;

but ▽ : ◖ :: △ : ● (hyp.),

∴ M ▽ : M ◡ :: ◇ : ● (B. 5. pr. 11);

and becaufe ◡ : ■ :: ◆ : ◇ (hyp.),

∴ M ◡ : *m* ■ :: ◆ : *m* ◇ (B. 5. pr. 4);

then, becaufe there are three magnitudes,

M ▽, M ◡, *m* ■,

and other three, M ◆, *m* ◇, *m* ●,

which, taken two and two in a crofs order, have

the fame ratio;

therefore, if M ▽ ⊏, =, or ⊐ *m* ■,

then will M ◆ ⊏, =, or ⊐ *m* ● (B. 5. pr. 21),

and ∴ ▽ : ■ :: ◆ : ● (B. 5. def. 5).

Next, let there be four magnitudes,

▽, ◡, ■, ◆,

and other four, ◇, ●, ▬, ▲,

which, when taken two and two in a crofs order, have

the fame ratio; namely,

▽ : ◡ :: ▬ : ▲,

◡ : ■ :: ● : ▬,

and ■ : ◆ :: ◇ : ●,

then fhall ▽ : ◆ :: ◇ : ▲.

For, becaufe ▽, ◡, ■ are three magnitudes,

and ⬤ , ■ , ▲ , other three,

which, taken two and two in a crofs order, have
the fame ratio,

therefore, by the firft cafe, ⬟ : ■ :: ⬤ : ▲ ,

but ■ : ◆ :: ◇ : ⬤ ,

therefore again, by the firft cafe, ⬟ : ◆ :: ◇ : ▲ ;

and fo on, whatever be the number of fuch magnitudes.

∴ If there be any number, &c.

F *the firſt has to the ſecond the ſame ratio which the third has to the fourth, and the fifth to the ſecond the ſame which the ſixth has to the fourth, the firſt and fifth together ſhall have to the ſecond the ſame ratio which the third and ſixth together have to the fourth.*

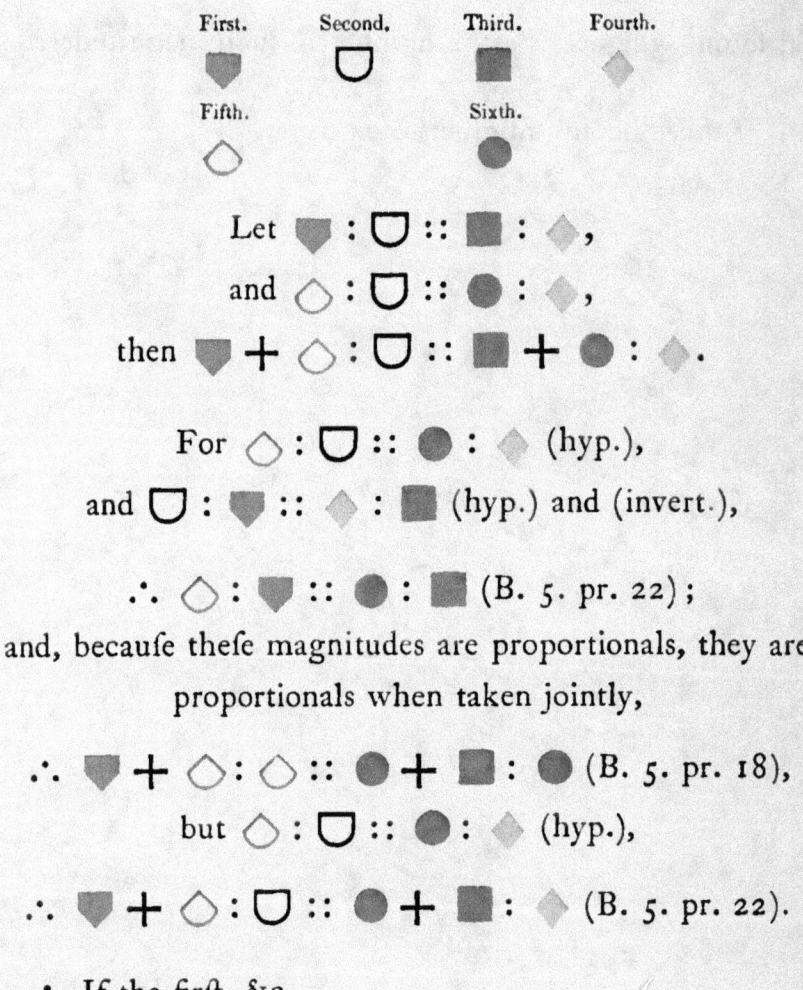

Let ◆ : ◔ :: ■ : ◆,

and ◇ : ◔ :: ● : ◆,

then ◆ + ◇ : ◔ :: ■ + ● : ◆.

For ◇ : ◔ :: ● : ◆ (hyp.),

and ◔ : ◆ :: ◆ : ■ (hyp.) and (invert.),

∴ ◇ : ◆ :: ● : ■ (B. 5. pr. 22);

and, becauſe theſe magnitudes are proportionals, they are proportionals when taken jointly,

∴ ◆ + ◇ : ◇ :: ● + ■ : ● (B. 5. pr. 18),

but ◇ : ◔ :: ● : ◆ (hyp.),

∴ ◆ + ◇ : ◔ :: ● + ■ : ◆ (B. 5. pr. 22).

∴ If the firſt, &c.

F *four magnitudes of the same kind are propor-tionals, the greateſt and leaſt of them together are greater than the other two together.*

Let four magnitudes, ▽ + ∪, ■ + ◆, ∪, and ◆,

of the ſame kind, be proportionals, that is to ſay,

$$▽ + ∪ : ■ + ◆ :: ∪ : ◆,$$

and let ▽ + ∪ be the greateſt of the four, and conſe-quently by pr. A and 14 of Book 5, ◆ is the leaſt;

then will ▽ + ∪ + ◆ be ⊏ ■ + ◆ + ∪;

becauſe ▽ + ∪ : ■ + ◆ :: ∪ : ◆,

∴ ▽ : ■ :: ▽ + ∪ : ■ + ◆ (B. 5. pr. 19),

but ▽ + ∪ ⊏ ■ + ◆ (hyp.),

∴ ▽ ⊏ ■ (B. 5. pr. A);

to each of theſe add ∪ + ◆,

∴ ▽ + ∪ + ◆ ⊏ ■ + ∪ + ◆.

∴ If four magnitudes, &c.

DEFINITION X.

WHEN three magnitudes are proportionals, the firſt is ſaid to have to the third the duplicate ratio of that which it has to the ſecond.

For example, if A, B, C, be continued proportionals, that is, $A : B :: B : C$, A is ſaid to have to C the duplicate ratio of $A : B$;

$$\text{or } \frac{A}{C} = \text{the ſquare of } \frac{A}{B}.$$

This property will be more readily ſeen of the quantities $a r^2$, ar, a, for $a r^2 : ar :: ar : a$;

$$\text{and } \frac{a r^2}{a} = r^2 = \text{the ſquare of } \frac{a r^2}{a r} = r,$$

$$\text{or of } a, \; a r, \; a r^2 ;$$

$$\text{for } \frac{a}{a r^2} = \frac{1}{r^2} = \text{the ſquare of } \frac{a}{a r} = \frac{1}{r}.$$

DEFINITION XI.

WHEN four magnitudes are continual proportionals, the firſt is ſaid to have to the fourth the triplicate ratio of that which it has to the ſecond; and ſo on, quadruplicate, &c. increaſing the denomination ſtill by unity, in any number of proportionals.

For example, let A, B, C, D, be four continued proportionals, that is, $A : B :: B : C :: C : D$; A is ſaid to have to D, the triplicate ratio of A to B ;

$$\text{or } \frac{A}{D} = \text{the cube of } \frac{A}{B}.$$

This definition will be better underftood, and applied to a greater number of magnitudes than four that are continued proportionals, as follows:—

Let ar^3, ar^2, ar, a, be four magnitudes in continued proportion, that is, $ar^3 : ar^2 :: ar^2 : ar :: ar : a,$

$$\text{then } \frac{ar^3}{a} = r^3 = \text{the cube of } \frac{ar^3}{ar^2} = r.$$

Or, let ar^5, ar^4, ar^3, ar^2, ar, a, be fix magnitudes in proportion, that is

$$ar^5 : ar^4 :: ar^4 \cdot ar^3 :: ar^3 : ar^2 :: ar^2 : ar :: ar : a,$$

then the ratio $\dfrac{ar^5}{a} = r^5 = $ the fifth power of $\dfrac{ar^5}{ar^4} = r.$

Or, let a, ar, ar^2, ar^3, ar^4, be five magnitudes in continued proportion; then $\dfrac{a}{ar^4} = \dfrac{1}{r^4} = $ the fourth power of $\dfrac{a}{ar} = \dfrac{1}{r}.$

DEFINITION A.

To know a compound ratio:—

When there are any number of magnitudes of the fame kind, the firft is faid to have to the laft of them the ratio compounded of the ratio which the firft has to the fecond, and of the ratio which the fecond has to the third, and of the ratio which the third has to the fourth; and fo on, unto the laft magnitude.

For example, if A, B, C, D, be four magnitudes of the fame kind, the firft A is faid to have to the laft D the ratio compounded of the ratio of A to B, and of the

A B C D
E F G H K L
M N

ratio of B to C, and of the ratio of C to D; or, the ratio of

DD

A to D is faid to be compounded of the ratios of A to B, B to C, and C to D.

And if A has to B the fame ratio which E has to F, and B to C the fame ratio that G has to H, and C to D the fame that K has to L ; then by this definition, A is faid to have to D the ratio compounded of ratios which are the fame with the ratios of E to F, G to H, and K to L. And the fame thing is to be underftood when it is more briefly expreffed by faying, A has to D the ratio compounded of the ratios of E to F, G to H, and K to L.

In like manner, the fame things being fuppofed ; if M has to N the fame ratio which A has to D, then for fhort-nefs fake, M is faid to have to N the ratio compounded of the ratios of E to F, G to H, and K to L.

This definition may be better underftood from an arith-metical or algebraical illuftration ; for, in fact, a ratio com-pounded of feveral other ratios, is nothing more than a ratio which has for its antecedent the continued product of all the antecedents of the ratios compounded, and for its confequent the continued product of all the confequents of the ratios compounded.

Thus, the ratio compounded of the ratios of

$$2 : 3, \ 4 : 7, \ 6 : 11, \ 2 : 5,$$

is the ratio of $2 \times 4 \times 6 \times 2 : 3 \times 7 \times 11 \times 5,$

or the ratio of $96 : 1155,$ or $32 : 385.$

And of the magnitudes A, B, C, D, E, F, of the fame kind, A : F is the ratio compounded of the ratios of

$$A : B, \ B : C, \ C : D, \ D : E, \ E : F;$$

for $A \times B \times C \times D \times E : B \times C \times D \times E \times F,$

or $\dfrac{A \times B \times C \times D \times E}{B \times C \times D \times E \times F} = \dfrac{A}{F},$ or the ratio of A : F.

ATIOS *which are compounded of the same ratios are the same to one another.*

Let A : B :: F : G,
B : C :: G : H,
C : D :: H : K,
and D : E :: K : L.

$$\begin{array}{ccccc} A & B & C & D & E \\ F & G & H & K & L \end{array}$$

Then the ratio which is compounded of the ratios of A : B, B : C, C : D, D : E, or the ratio of A : E, is the same as the ratio compounded of the ratios of F : G, G : H, H : K, K : L, or the ratio of F : L.

For $\frac{A}{B} = \frac{F}{G}$,

$$\frac{B}{C} = \frac{G}{H},$$

$$\frac{C}{D} = \frac{H}{K},$$

and $\frac{D}{E} = \frac{K}{L}$;

$$\therefore \frac{A \times B \times C \times D}{B \times C \times D \times E} = \frac{F \times G \times H \times K}{G \times H \times K \times L},$$

and $\therefore \frac{A}{E} = \frac{F}{L}$,

or the ratio of A : E is the same as the ratio of F : L.

The same may be demonstrated of any number of ratios so circumstanced.

Next, let A : B :: K : L,
B : C :: H : K,
C : D :: G : H,
D : E :: F : G.

Then the ratio which is compounded of the ratios of
A : B, B : C, C : D, D : E, or the ratio of A : E, is the
fame as the ratio compounded of the ratios of E : L, H : K,
G : H, F : G, or the ratio of F : L.

$$\text{For } \frac{A}{B} = \frac{K}{L},$$

$$\frac{B}{C} = \frac{H}{K},$$

$$\frac{C}{D} = \frac{G}{H},$$

$$\text{and } \frac{D}{E} = \frac{F}{G};$$

$$\therefore \frac{A \times B \times C \times D}{B \times C \times D \times E} = \frac{K \times H \times G \times F}{L \times K \times H \times G},$$

$$\text{and } \therefore \frac{A}{E} = \frac{F}{L},$$

or the ratio of A : E is the fame as the ratio of F : L.

∴ Ratios which are compounded, &c.

F *several ratios be the same to several ratios, each to each, the ratio which is compounded of ratios which are the same to the first ratios, each to each, shall be the same to the ratio compounded of ratios which are the same to the other ratios, each to each.*

$$
\begin{array}{ccccccc cccc}
A & B & C & D & E & F & G & H & & P & Q & R & S & T \\
a & b & c & d & e & f & g & h & & V & W & X & Y & Z
\end{array}
$$

If $A : B :: a : b$ and $A : B :: P : Q$ $a : b :: V : W$

$\quad C : D :: c : d$ $C : D :: Q : R$ $c : d :: W : X$

$\quad E : F :: e : f$ $E : F :: R : S$ $e : f :: X : Y$

and $G : H :: g : h$ $G : H :: S : T$ $g : h :: Y : Z$

then $P : T = V : Z.$

For $\dfrac{P}{Q} = \dfrac{A}{B} = \dfrac{a}{b} = \dfrac{V}{W},$

$\quad\ \dfrac{Q}{R} = \dfrac{C}{D} = \dfrac{c}{d} = \dfrac{W}{X},$

$\quad\ \dfrac{R}{S} = \dfrac{E}{F} = \dfrac{e}{f} = \dfrac{X}{Y},$

$\quad\ \dfrac{S}{T} = \dfrac{G}{H} = \dfrac{g}{h} = \dfrac{Y}{Z};$

and $\therefore \dfrac{P \times Q \times R \times S}{Q \times R \times S \times T} = \dfrac{V \times W \times X \times Y}{W \times X \times Y \times Z},$

and $\therefore \dfrac{P}{T} = \dfrac{V}{Z},$

or $P : T = V : Z.$

$\therefore$ If several ratios, &c.

F *a ratio which is compounded of several ratios be the same to a ratio which is compounded of several other ratios ; and if one of the first ratios, or the ratio which is compounded of several of them, be the same to one of the last ratios, or to the ratio which is compounded of several of them ; then the remaining ratio of the first, or, if there be more than one, the ratio compounded of the remaining ratios, shall be the same to the remaining ratio of the last, or, if there be more than one, to the ratio compounded of these remaining ratios.*

$$\boxed{\begin{array}{l} \text{A B C D E F G H} \\ \text{P Q R S T X} \end{array}}$$

Let A : B, B : C, C : D, D : E, E : F, F : G, G : H, be the first ratios, and P : Q, Q : R, R : S, S : T, T : X, the other ratios ; also, let A : H, which is compounded of the first ratios, be the same as the ratio of P : X, which is the ratio compounded of the other ratios ; and, let the ratio of A : E, which is compounded of the ratios of A : B, B : C, C : D, D : E, be the same as the ratio of P : R, which is compounded of the ratios P : Q, Q : R.

Then the ratio which is compounded of the remaining first ratios, that is, the ratio compounded of the ratios E : F, F : G, G : H, that is, the ratio of E : H, shall be the same as the ratio of R : X, which is compounded of the ratios of R : S, S : T, T : X, the remaining other ratios.

Becaufe $\dfrac{A \times B \times C \times D \times E \times F \times G}{B \times C \times D \times E \times F \times G \times H} = \dfrac{P \times Q \times R \times S \times T}{Q \times R \times S \times T \times X}$,

or $\dfrac{A \times B \times C \times D}{B \times C \times D \times E} \times \dfrac{E \times F \times G}{F \times G \times H} = \dfrac{P \times Q}{Q \times R} \times \dfrac{R \times S \times T}{S \times T \times X}$,

and $\dfrac{A \times B \times C \times D}{B \times C \times D \times E} = \dfrac{P \times Q}{Q \times R}$,

$\therefore \dfrac{E \times F \times G}{F \times G \times H} = \dfrac{R \times S \times T}{S \times T \times X}$,

$\therefore \dfrac{E}{H} = \dfrac{R}{X}$,

$\therefore E : H = R : X.$

$\therefore$ If a ratio which, &c.

F *there be any number of ratios, and any number of other ratios, such that the ratio which is compounded of ratios, which are the same to the first ratios, each to each, is the same to the ratio which is compounded of ratios, which are the same, each to each, to the last ratios—and if one of the first ratios, or the ratio which is compounded of ratios, which are the same to several of the first ratios, each to each, be the same to one of the last ratios, or to the ratio which is compounded of ratios, which are the same, each to each, to several of the last ratios—then the remaining ratio of the first; or, if there be more than one, the ratio which is compounded of ratios, which are the same, each to each, to the remaining ratios of the first, shall be the same to the remaining ratio of the last; or, if there be more than one, to the ratio which is compounded of ratios, which are the same, each to each, to these remaining ratios.*

$$
\begin{array}{l}
\quad\quad\quad \text{h \ k\ m\ n\ s} \\
\text{AB, CD, EF, GH, KL, MN,} \quad a\ b\ c\ d\ e\ f\ g \\
\text{OP, QR, ST, VW, XY,} \quad\quad h\ k\ l\ m\ n\ p \\
\quad\quad \text{a\ b\ c\ d} \quad\quad \text{e\ f\ g}
\end{array}
$$

Let A:B, C:D, E:F, G:H, K:L, M:N, be the first ratios, and O:P, Q:R, S:T, V:W, X:Y, the other ratios;

$$
\begin{aligned}
\text{and let } A:B &= a:b, \\
C:D &= b:c, \\
E:F &= c:d, \\
G:H &= d:e, \\
K:L &= e:f, \\
M:N &= f:g.
\end{aligned}
$$

Then, by the definition of a compound ratio, the ratio of $a:g$ is compounded of the ratios of $a:b$, $b:c$, $c:d$, $d:e$, $e:f$, $f:g$, which are the fame as the ratio of $A:B$, $C:D$, $E:F$, $G:H$, $K:L$, $M:N$, each to each.

$$
\begin{aligned}
\text{Alfo, } O:P &= h:k, \\
Q:R &= k:l, \\
S:T &= l:m, \\
V:W &= m:n, \\
X:Y &= n:p.
\end{aligned}
$$

Then will the ratio of $h:p$ be the ratio compounded of the ratios of $h:k$, $k:l$, $l:m$, $m:n$, $n:p$, which are the fame as the ratios of $O:P$, $Q:R$, $S:T$, $V:W$, $X:Y$, each to each.

$\therefore$ by the hypothefis $a:g = h:p$.

Alfo, let the ratio which is compounded of the ratios of $A:B$, $C:D$, two of the firft ratios (or the ratios of $a:c$, for $A:B = a:b$, and $C:D = b:c$), be the fame as the ratio of $a:d$, which is compounded of the ratios of $a:b$, $b:c$, $c:d$, which are the fame as the ratios of $O:P$, $Q:R$, $S:T$, three of the other ratios.

And let the ratios of $h:s$, which is compounded of the ratios of $h:k$, $k:m$, $m:n$, $n:s$, which are the fame as the remaining firft ratios, namely, $E:F$, $G:H$, $K:L$, $M:N$; alfo, let the ratio of $e:g$, be that which is compounded of the ratios $e:f$, $f:g$, which are the fame, each to each, to the remaining other ratios, namely, $V:W$, $X:Y$. Then the ratio of $h:s$ fhall be the fame as the ratio of $e:g$; or $h:s = e:g$.

For $\dfrac{A \times C \times E \times G \times K \times M}{B \times D \times F \times H \times L \times N} = \dfrac{a \times b \times c \times d \times e \times f}{b \times c \times d \times e \times f \times g}$,

E E

and $\dfrac{O \times Q \times S \times V \times X}{P \times R \times T \times W \times Y} = \dfrac{h \times k \times l \times m \times n}{k \times l \times m \times n \times p}$,

by the compofition of the ratios;

$\therefore \dfrac{a \times b \times c \times d \times e \times f}{b \times c \times d \times e \times f \times g} = \dfrac{h \times k \times l \times m \times n}{k \times l \times m \times n \times p}$ (hyp.),

or $\dfrac{a \times b}{b \times c} \times \dfrac{c \times d \times e \times f}{d \times e \times f \times g} = \dfrac{h \times k \times l}{k \times l \times m} \times \dfrac{m \times n}{n \times p}$,

but $\dfrac{a \times b}{b \times c} = \dfrac{A \times C}{B \times D} = \dfrac{O \times Q \times S}{P \times R \times T} = \dfrac{a \times b \times c}{b \times c \times d} = \dfrac{h \times k \times l}{k \times l \times m}$;

$\therefore \dfrac{c \times d \times e \times f}{d \times e \times f \times g} = \dfrac{m \times n}{n \times p}$.

And $\dfrac{c \times d \times e \times f}{d \times e \times f \times g} = \dfrac{h \times k \times m \times n}{k \times m \times n \times s}$ (hyp.),

and $\dfrac{m \times n}{n \times p} = \dfrac{e \times f}{f \times g}$ (hyp.),

$\therefore \dfrac{h \times k \times m \times n}{k \times m \times n \times s} = \dfrac{e\,f}{f\,g}$,

$\therefore \dfrac{h}{s} = \dfrac{e}{g}$,

$\therefore h : s = e : g.$

$\therefore$ If there be any number, &c.

₂ Algebraical and Arithmetical expositions of the Fifth Book of Euclid are given in Byrne's Doctrine of Proportion; published by WILLIAMS and Co. London. 1841.

BOOK VI.
DEFINITIONS.

I.

ECTILINEAR figures are said to be similar, when they have their several angles equal, each to each, and the sides about the equal angles proportional.

II.

Two sides of one figure are said to be reciprocally proportional to two sides of another figure when one of the sides of the first is to the second, as the remaining side of the second is to the remaining side of the first.

III.

A straight line is said to be cut in extreme and mean ratio, when the whole is to the greater segment, as the greater segment is to the less.

IV.

The altitude of any figure is the straight line drawn from its vertex perpendicular to its base, or the base produced.

RIANGLES and parallelograms having the same altitude are to one another as their bases.

Let the triangles ▲ and ▲ have a common vertex, and their bases ━━ and ━

in the same straight line.

Produce ━━━━ both ways, take succeffively on ━━ produced lines equal to it; and on ━━━ produced lines succeffively equal to it; and draw lines from the common vertex to their extremities.

The triangles ◢ thus formed are all equal to one another, fince their bafes are equal. (B. 1. pr. 38.)

∴ ◢ and its bafe are refpectively equi-

multiples of ◢ and the bafe ━━ .

212

In like manner  and its bafe are refpec-

tively equimultiples of ▲ and the bafe ▬ .

∴ If *m* or 6 times ▋ ⊏ ═ or ⊐ *n* or 5 times ▊
then *m* or 6 times ▬ ⊏ ═ or ⊐ *n* or 5 times ▬ ,
m and *n* ftand for every multiple taken as in the fifth
definition of the Fifth Book. Although we have only
fhown that this property exifts when *m* equal 6, and *n*
equal 5, yet it is evident that the property holds good for
every multiple value that may be given to *m*, and to *n*.

∴ ◢ : ◣ :: ▬ : ▬ (B. 5. def. 5.)

Parallelograms having the fame altitude are the doubles
of the triangles, on their bafes, and are proportional to
them (Part 1), and hence their doubles, the parallelograms,
are as their bafes. (B. 5. pr. 15.)

Q. E. D.

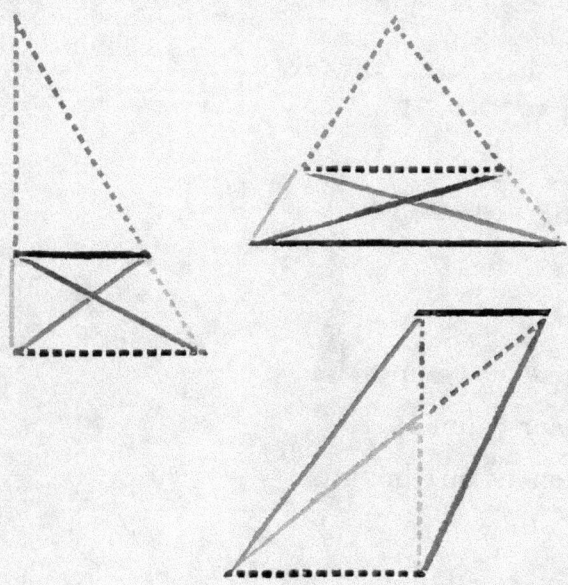

F *a ftraight line* ——— *be drawn parallel to any fide* ••••••••• *of a tri-angle, it fhall cut the other fides, or thofe fides produced, into pro-portional fegments.*

And if any ftraight line ——— *divide the fides of a triangle, or thofe fides produced, into proportional feg-ments, it is parallel to the remaining fide* ••••••••• .

PART I.

Let ——— ‖ ••••••••• , then fhall

——— : ••••••••• :: •••••••• : •••••••• .

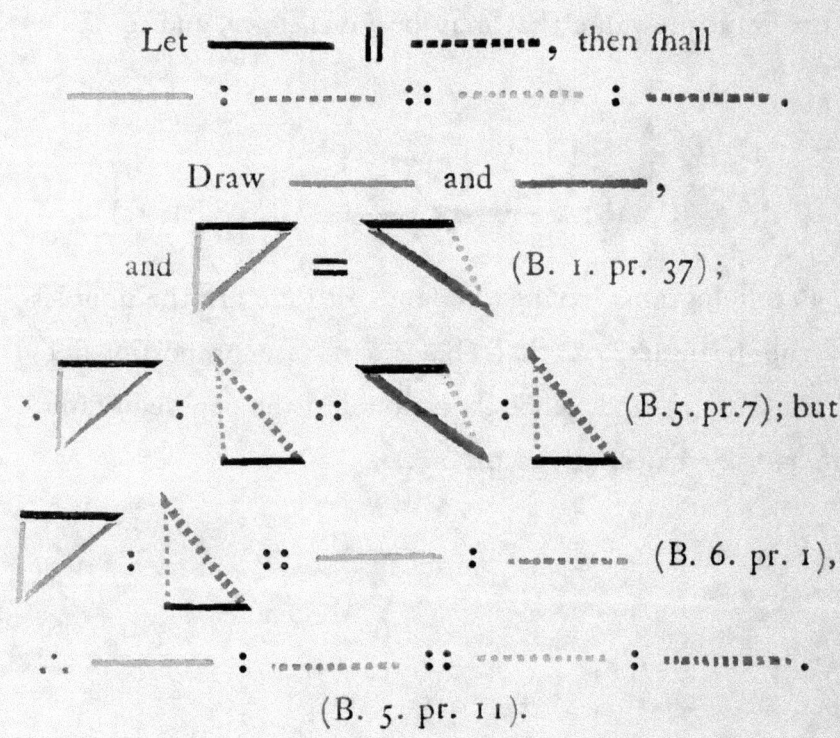

Draw ——— and ——— ,

and ◥ = ◣   (B. 1. pr. 37);

◥ : ◣ :: ◣ : ◣   (B. 5. pr. 7); but

◥ : ◣ :: ——— : •••••••••   (B. 6. pr. 1),

∴ ——— : ••••••••• :: •••••••• : •••••••• .

(B. 5. pr. 11).

PART II.

Let ——— : ••••••••• :: ••••••••• : ••••••••• ,

then ——— || ••••••••• .

Let the fame conftruction remain,

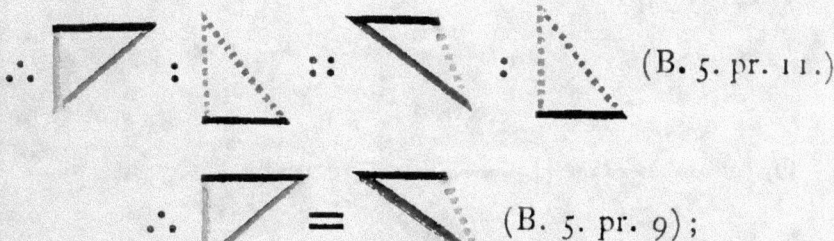

becaufe ——— : ••••••••• :: ◥ : ◿ }

and ••••••••• : ••••••••• :: ◥ : ◿ } (B. 6. pr. 1);

but ——— : ••••••••• :: ••••••••• : ••••••••• (hyp.),

∴ ◥ : ◿ :: ◥ : ◿ (B. 5. pr. 11.)

∴ ◥ = ◸ (B. 5. pr. 9);

but they are on the fame bafe ••••••••• , and at the
fame fide of it, and

∴ ——— || ••••••••• (B. 1. pr. 39).

Q. E. D.

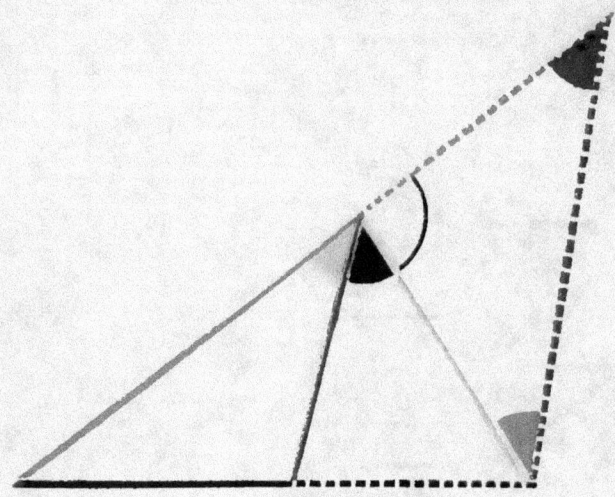

A RIGHT *line* (——) *bisecting the angle of a triangle, divides the opposite side into segments* (——, - - - - -) *proportional to the conterminous sides* (——, ——).

And if a straight line (——) *drawn from any angle of a triangle divide the opposite side* (—— · · ·) *into segments* (——, · · · · ·) *proportional to the conterminous sides* (——, ——), *it bisects the angle.*

PART I.

Draw · · · · · || ——, to meet · · · · · ;

then, ◄ = ◄   (B. 1. pr. 29),

∴ ◄ = ◄ ; but ◄ = ◄, ∴ ◄ = ◄,

∴ · · · · · = ——   (B. 1. pr. 6);

and because —— || · · · · · ,

· · · · · : —— :: · · · · · : ——

(B. 6. pr. 2);

but · · · · · = —— ;

∴ —— : —— :: · · · · · : ——

(B. 5. pr. 7).

PART II.

Let the fame conftruction remain,

and ▬▬▬ : ·········· :: ▬▬▬ : ········· (B. 6. pr. 2) ;

but ▬▬▬ : ·········· :: ▬▬▬ : ▬▬▬ (hyp.)

∴ ▬▬▬ : ·········· :: ▬▬▬ : ▬▬▬ (B. 5. pr. 11).

and ∴ ·········· ═ ▬▬▬ (B. 5. pr. 9),

and ∴ ◀ ═ ◀ (B. 1. pr. 5) ; but fince

▬▬▬ ‖ ·········· ; ▲ ═ ◀,

and ◀ ═ ◀ (B. 1. pr. 29) ;

∴ ◀ ═ ◀, and ◀ ═ ▲,

and ∴ ▬▬▬ bifects ▲.

Q. E. D.

F F

217

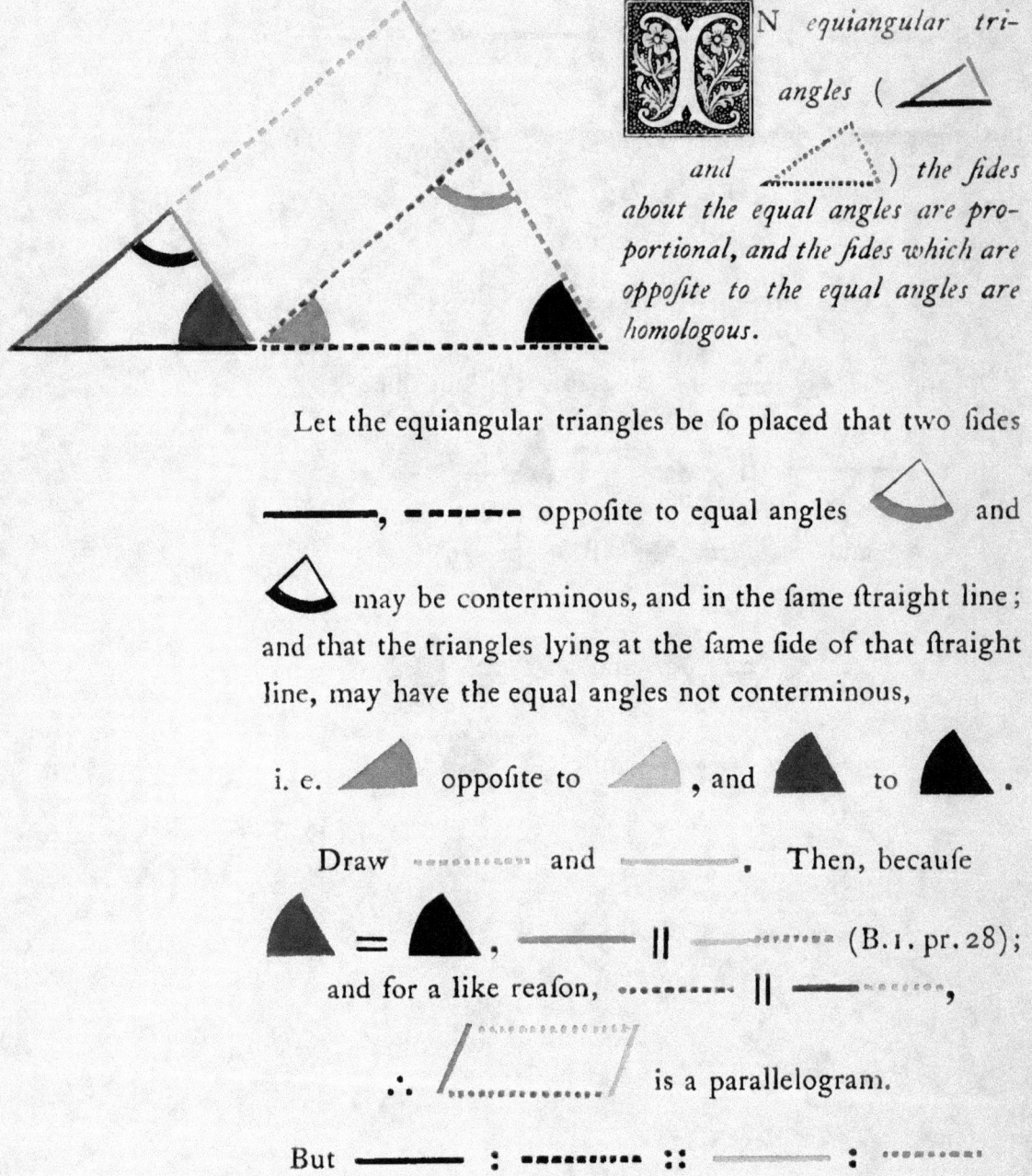

N equiangular triangles (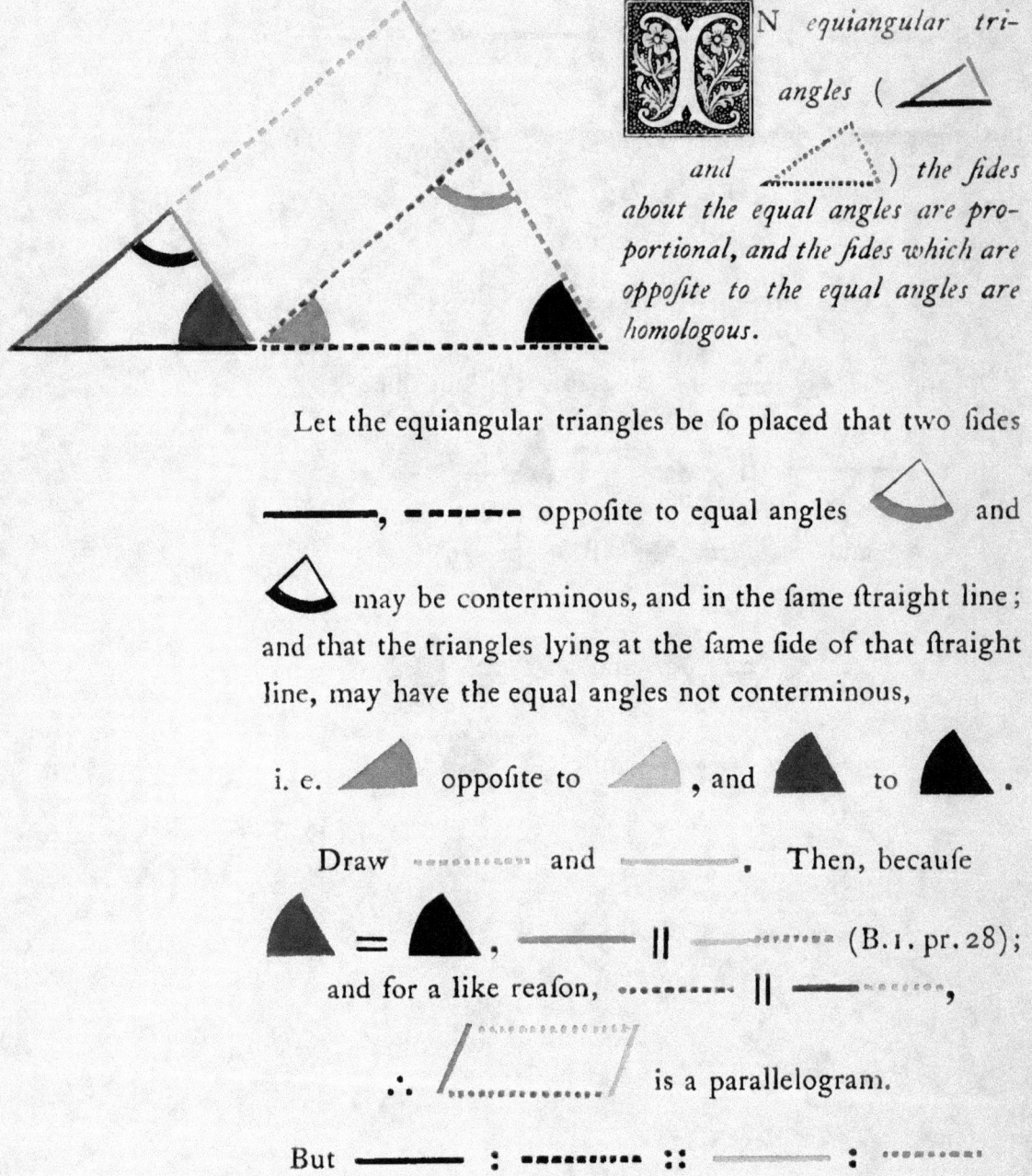 and 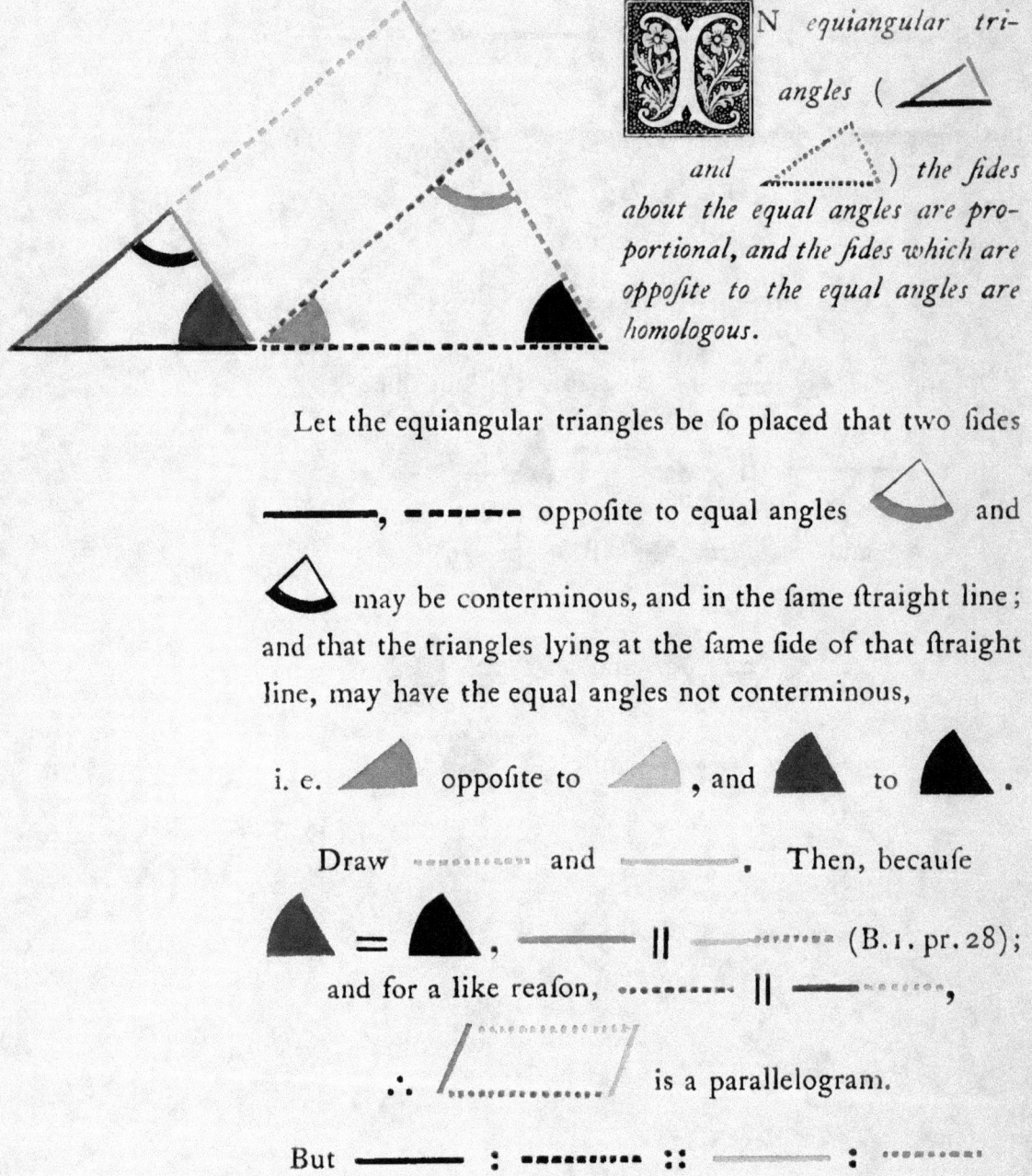) the *fides* about the equal angles are proportional, and the *fides* which are oppofite to the equal angles are homologous.

Let the equiangular triangles be fo placed that two fides

—— , - - - - - oppofite to equal angles and

 may be conterminous, and in the fame ftraight line; and that the triangles lying at the fame fide of that ftraight line, may have the equal angles not conterminous,

i. e. oppofite to , and to .

Draw ———— and ———— . Then, becaufe

 = , —— || ——— (B. 1. pr. 28);

and for a like reafon, ·········· || ———·····,

∴ ———— is a parallelogram.

But —— : ·········· :: ———— : ··········
(B. 6. pr. 2);

and since ——— = ——— (B. 1. pr. 34),

——— : ·········· :: ——— : ·········· ; and by

alternation, ——— : ——— :: ·········· : ··········

(B. 5. pr. 16).

In like manner it may be fhown, that

——— : ·········· :: ——— : ·········· ;

and by alternation, that

——— : ——— :: ·········· : ·········· ;

but it has been already proved that

——— : ——— :: ·········· : ·········· ,

and therefore, ex æquali,

——— : ——— :: ·········· : ··········

(B. 5. pr. 22),

therefore the fides about the equal angles are proportional,
and thofe which are oppofite to the equal angles
are homologous.

Q. E. D.

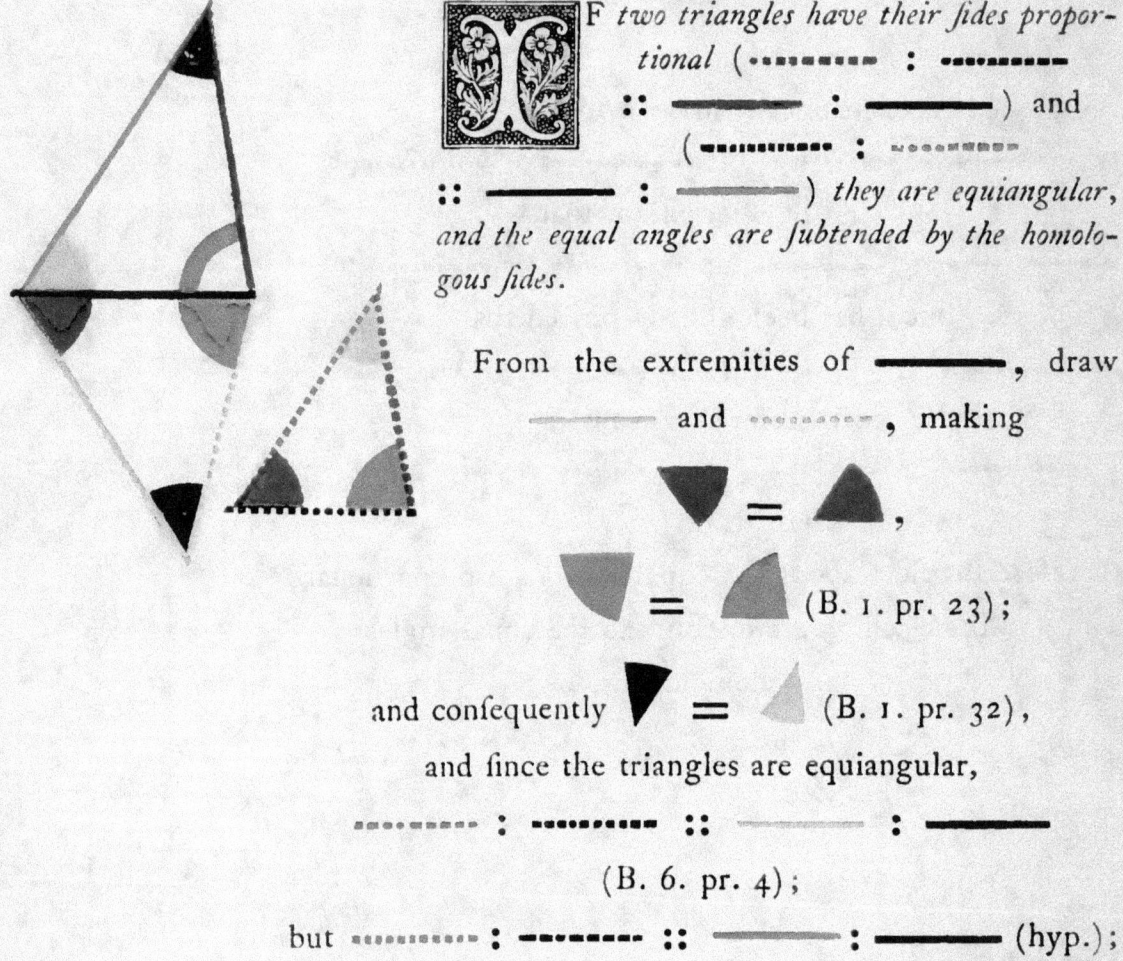

F two triangles have their sides propor-
tional (............... : .... :: ——————— : ———————) and
(............... : .... :: ——————— : ———————) *they are equiangular,
and the equal angles are subtended by the homolo-
gous sides.*

From the extremities of ———————, draw
——————— and ...., making

▼ = ▲ ,

◣ = ◥ (B. 1. pr. 23) ;

and consequently ▼ = ◣ (B. 1. pr. 32),
and since the triangles are equiangular,

.... : .... :: ——————— : ———————

(B. 6. pr. 4) ;

but .... : .... :: ——————— : ——————— (hyp.);

∴ ——————— : ——————— :: ——————— : ———————,

and consequently ——————— = ——————— (B. 5. pr. 9).

In the like manner it may be shown that

——————— = ...

Therefore, the two triangles having a common bafe ———, and their fides equal, have alfo equal angles oppofite to equal fides, i. e.

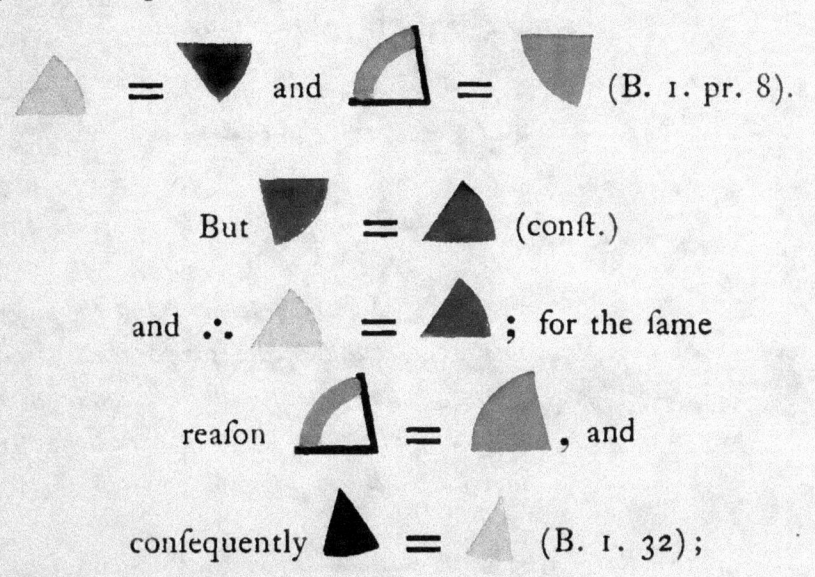

and therefore the triangles are equiangular, and it is evident that the homologous fides fubtend the equal angles.

Q. E. D.

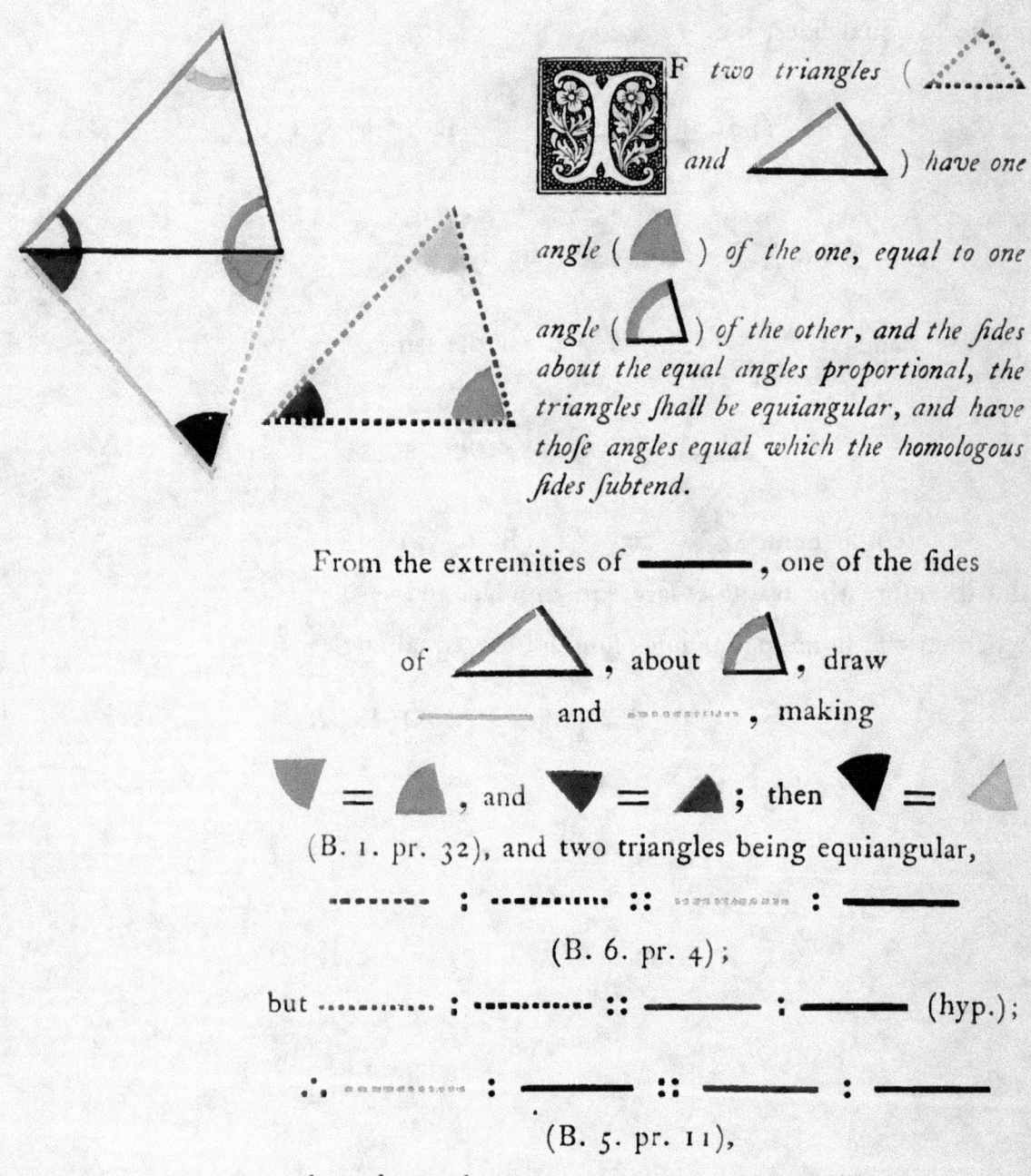

F *two triangles (*) *and*) *have one*

angle ( ) *of the one, equal to one*

angle ( ) *of the other, and the fides about the equal angles proportional, the triangles fhall be equiangular, and have thofe angles equal which the homologous fides fubtend.*

From the extremities of ——————, one of the fides

of , about , draw

—————— and , making

= , and = ; then =

(B. 1. pr. 32), and two triangles being equiangular,

: :: :

(B. 6. pr. 4);

but : :: : (hyp.);

∴ : :: : (B. 5. pr. 11),

and confequently = (B. 5. pr. 9);

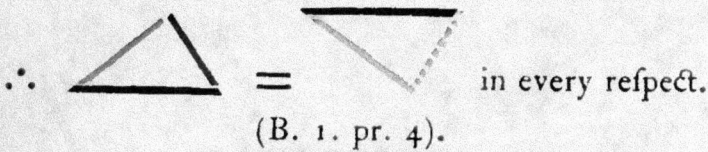 in every refpect.

(B. 1. pr. 4).

But ▼ = ▲ (conft.),

and ∴ △ = ▲ ; and

fince alfo △ = ◢ ,

△ = ◢ (B. 1. pr. 32);

and ∴ ⟁ and ◺ are equiangular, with their equal angles oppofite to homologous fides.

Q. E. D.

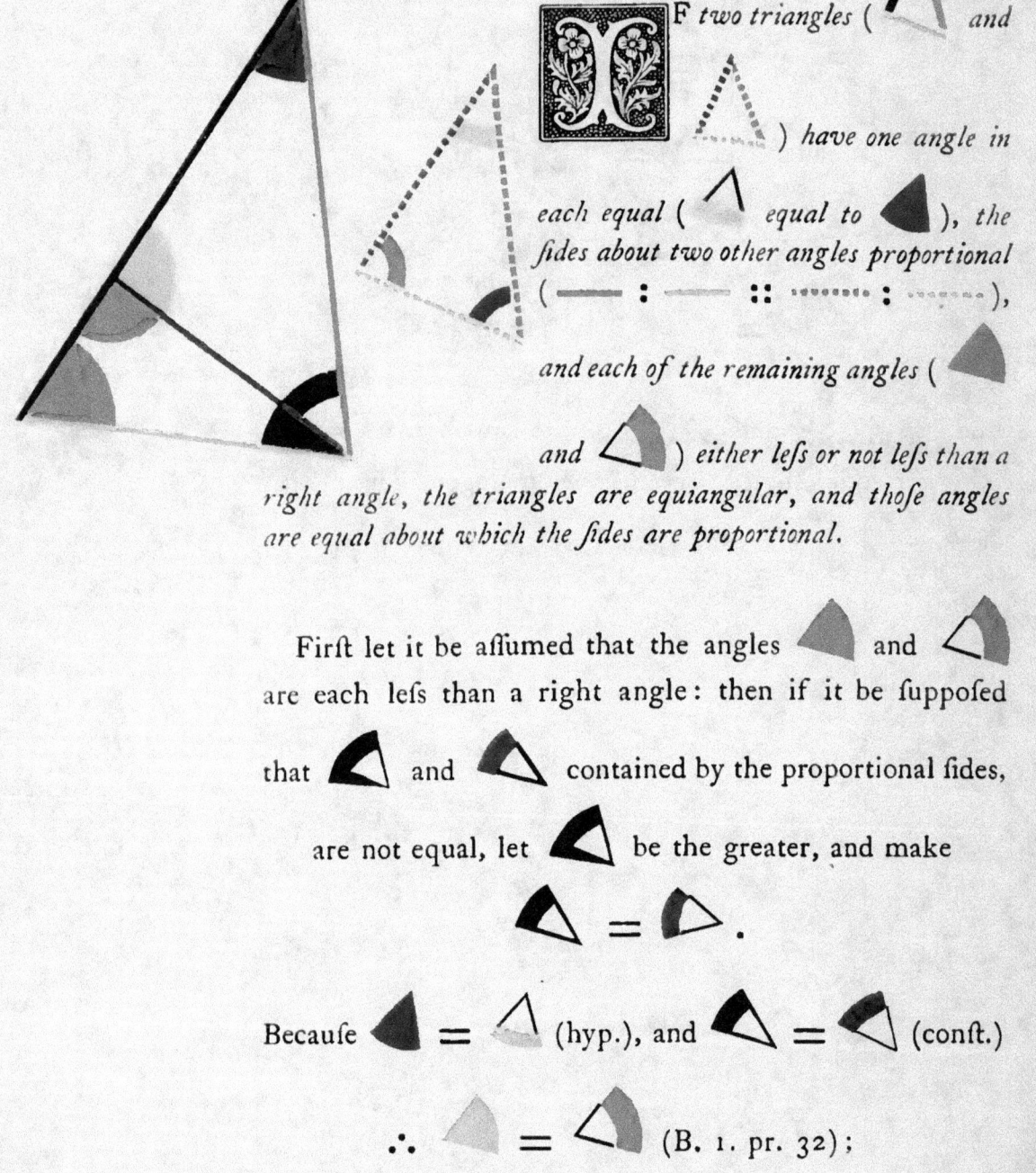

F *two triangles* (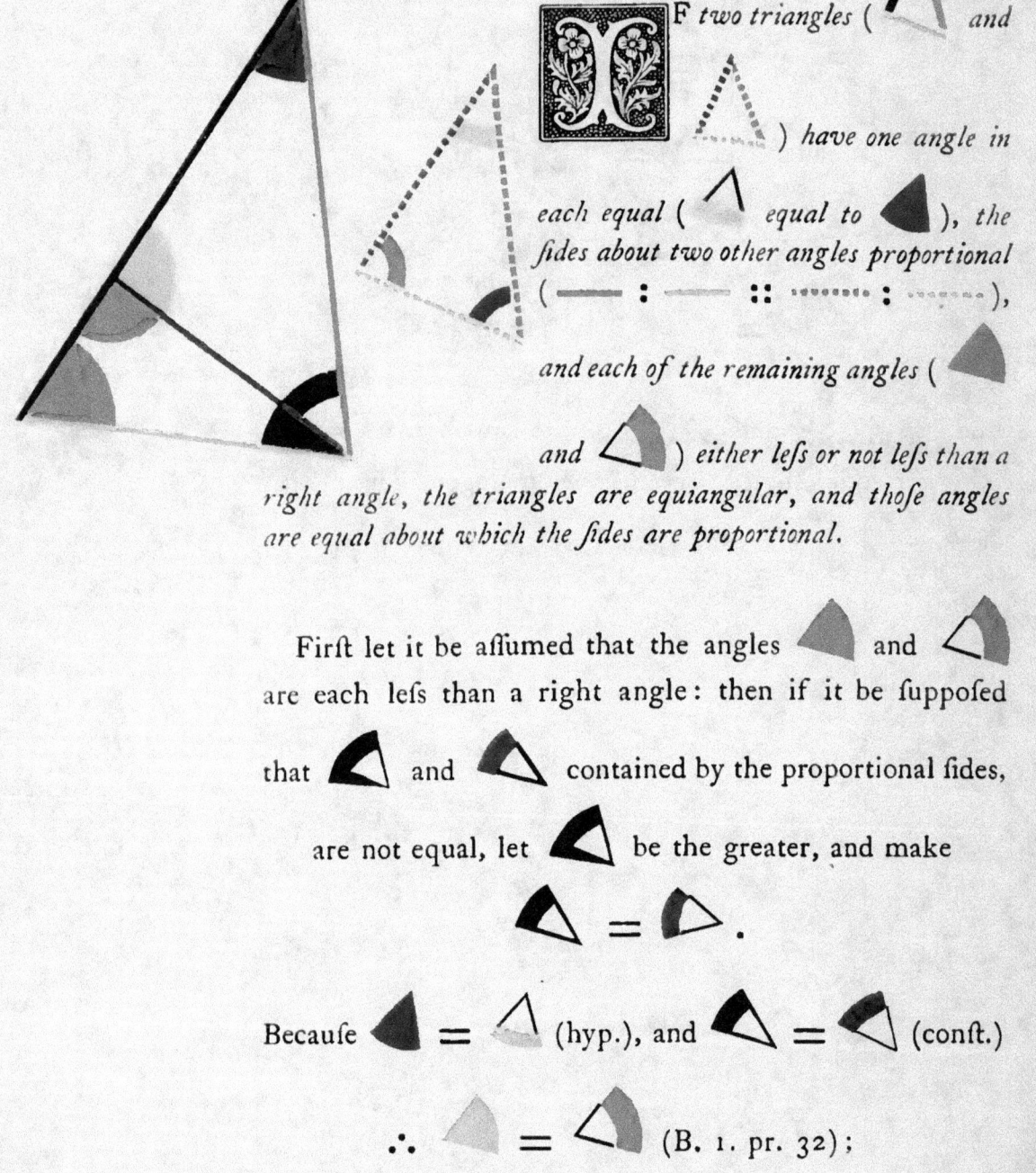 *and*) *have one angle in each equal* (*equal to*), *the sides about two other angles proportional* (— : — :: ⋯⋯ : ⋯⋯),

and each of the remaining angles (

and) *either less or not less than a right angle, the triangles are equiangular, and those angles are equal about which the sides are proportional.*

First let it be affumed that the angles and are each lefs than a right angle: then if it be fuppofed

that and contained by the proportional fides,

are not equal, let be the greater, and make

$$ \text{\includegraphics{}} = \text{\includegraphics{}} . $$

Becaufe = (hyp.), and = (conft.)

∴ = (B. 1. pr. 32);

∴ ——— : ——— ∷ ········ : ═══ (B. 6. pr. 4),

but ——— : ═══ ∷ ········ : ═══ (hyp.)

∴ ——— : ——— ∷ ——— : ——— ;

∴ ——— ═ ——— (B. 5. pr. 9),

and ∴ ◢ ═ ◣ (B. 1. pr. 5).

But ◢ is lefs than a right angle (hyp.)

∴ ◢ is lefs than a right angle; and ∴ ◣ muſt be greater than a right angle (B. 1. pr. 13), but it has been proved ═ ◁ and therefore lefs than a right angle, which is abſurd. ∴ ◣ and ◺ are not unequal;

∴ they are equal, and ſince ◣ ═ ◺ (hyp.)

∴ ◢ ═ ◁ (B. 1. pr. 32), and therefore the triangles are equiangular.

But if ◢ and ◺ be aſſumed to be each not lefs than a right angle, it may be proved as before, that the triangles are equiangular, and have the ſides about the equal angles proportional. (B. 6. pr. 4).

Q. E. D.

G G

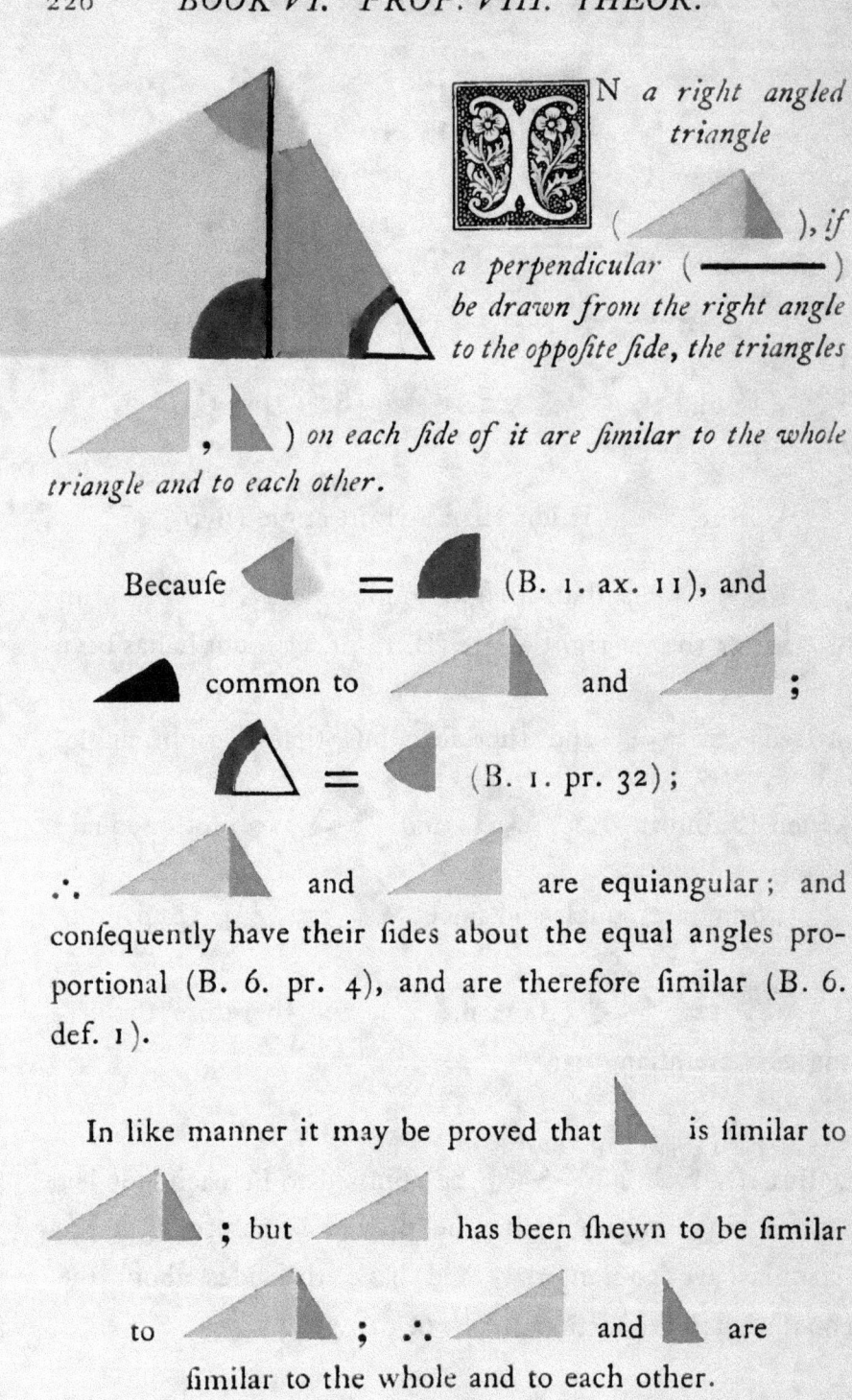

IN a right angled triangle ( ), if a perpendicular (————) be drawn from the right angle to the oppofite fide, the triangles (,) on each fide of it are fimilar to the whole triangle and to each other.

Becaufe = (B. 1. ax. 11), and common to and ; = (B. 1. pr. 32);

∴ and are equiangular; and confequently have their fides about the equal angles proportional (B. 6. pr. 4), and are therefore fimilar (B. 6. def. 1).

In like manner it may be proved that is fimilar to ; but has been fhewn to be fimilar to ; ∴ and are fimilar to the whole and to each other.

Q. E. D.

ROM *a given straight line* (———) *to cut off any required part.*

From either extremity of the given line draw ——— making any angle with ———; and produce ——— till the whole produced line ——— contains ——— as often as ——— contains the required part.

Draw ———, and draw ——— ‖ ———.

——— is the required part of ———.

For since ——— ‖ ———

——— : ——— :: ——— : ———

(B. 6. pr. 2), and by compofition (B. 5. pr. 18);

——— : ——— :: ——— : ———;

but ——— contains ——— as often as ——— contains the required part (conſt.);

∴ ——— is the required part.

Q. E. D.

227

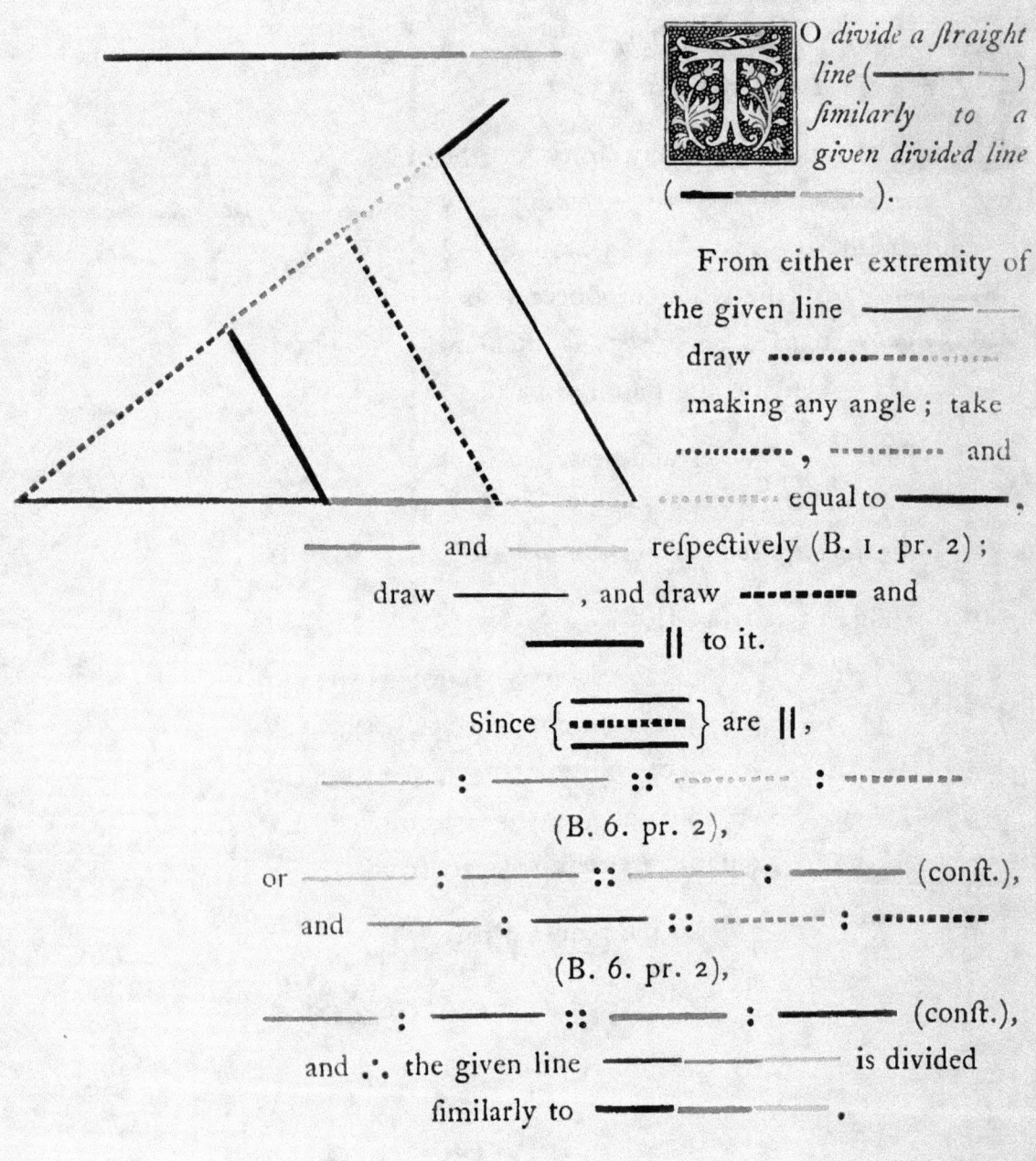

O *divide a straight line* (————) *similarly to a given divided line* ( ).

From either extremity of the given line ————— draw •••••••••—•• making any angle; take ••••••••• , —————— and ———————— equal to ——————— , respectively (B. 1. pr. 2); draw ——————— , and draw ••••••••••• and ━━━━━━ ‖ to it.

Since ⎰ ••••••••• ⎱ are ‖ ,
      ⎱ —————— ⎰

———— : ————— :: •••••••• : ••••••••
(B. 6. pr. 2),

or ———— : ———— :: ——————— : ——————— (conſt.),

and ———— : ——— :: ••••••••• : •••••••
(B. 6. pr. 2),

———— : ——— :: ———— : ———— (conſt.),

and ∴ the given line ——————— is divided ſimilarly to ———————— .

Q. E. D.

T O *find a third proportional to two given ſtraight lines* (——— *and* ———).

At either extremity of the given line ——— draw ------——— making an angle; take ------ = ———, and draw ——— ; make ------ = ———, and draw --------- || ———; (B. 1. pr. 31.)

——— is the third proportional to ——— and ———.

For ſince ——— || ---------,

∴ ——— : ------ :: ------ : ——— (B. 6 pr. 2);

but ------ = --------- = ——— (conſt.);

∴ ——— : ------ :: ——— : ——— . (B. 5. pr. 7).

Q. E. D.

O *find a fourth pro-*
portional to three
given lines

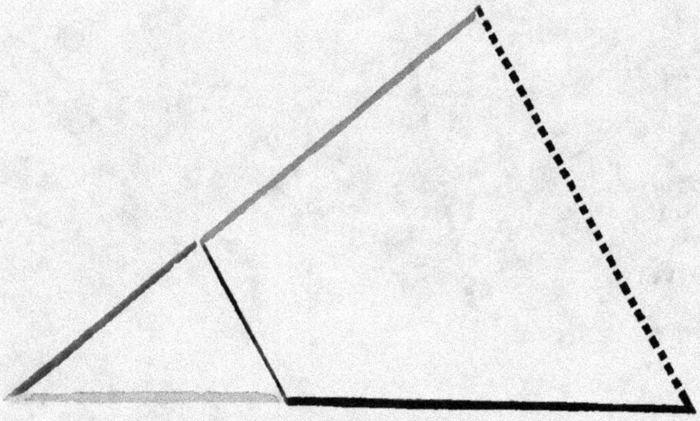

$$\left\{ \begin{matrix} \rule{3cm}{0.4mm} \\ \rule{3cm}{0.4mm} \\ \rule{3cm}{0.4mm} \end{matrix} \right\}.$$

Draw ▬▬▬▬▬

and ━━━━━ making any angle;

take ━━━━ = ━━━━,

and ━━━━ = ━━━━,

also ━━━━ = ━━━━,

draw ─────,

and ━━━━ ‖ ─────;

(B. 1. pr. 31);

━━━━ is the fourth proportional.

On account of the parallels,

━━━ : ━━━ :: ━━━ : ━━━

(B. 6. pr. 2);

but $\left\{ \begin{matrix} \rule{2cm}{0.3mm} \\ \rule{2cm}{0.3mm} \\ \rule{2cm}{0.3mm} \end{matrix} \right\} = \left\{ \begin{matrix} \rule{2cm}{0.4mm} \\ \rule{2cm}{0.4mm} \\ \rule{2cm}{0.4mm} \end{matrix} \right\}$ (conſt.);

∴ ━━━ : ━━━ :: ━━━ : ━━━.

(B. 5. pr. 7).

Q. E. D.

O *find a mean propor-*
tional between two given
straight lines

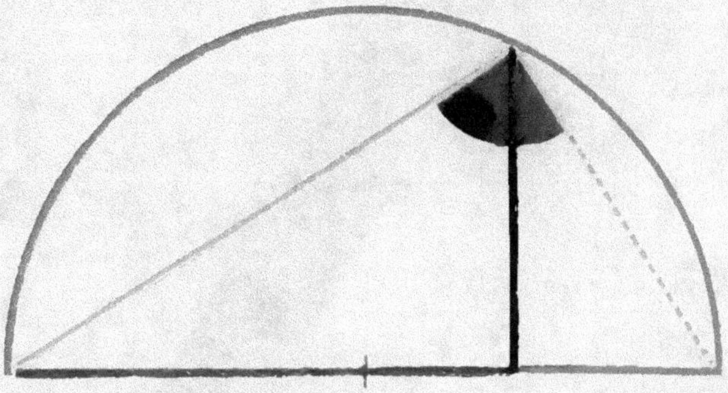

{ ⸺⸺ }.

Draw any ftraight line ⸺⸺,

make ⸺ = ⸺⸺,

and ⸺⸺ = ⸺⸺; bifect ⸺⸺ :

and from the point of bifection as a centre, and half the

line as a radius, defcribe a femicircle ⌒.

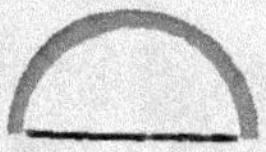

draw ⸺ ⊥ ⸺ :

⸺⸺ is the mean proportional required.

Draw ⸺⸺ and ⸺⸺.

Since ◣ is a right angle (B. 3. pr. 31),

and ⸺⸺ is ⊥ from it upon the oppofite fide,

∴ ⸺⸺ is a mean proportional between

⸺⸺ and ⸺⸺ (B. 6. pr. 8),

and ∴ between ⸺⸺ and ⸺⸺ (conft.).

Q. E. D

I.

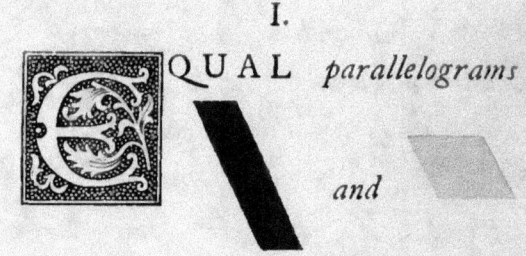

QUAL *parallelograms* and ,

which have one angle in each equal, have the fides about the equal angles reciprocally proportional

(⸺ : ⸺ :: ⸺ : ⸺).

II.

And parallelograms which have one angle in each equal, and the fides about them reciprocally proportional, are equal.

Let ⸺ and ⸺ ; and ⸺ and ⸺ , be fo placed that ⸺ ⸺ and ⸺ ⸺ may be continued right lines. It is evident that they may affume this pofition. (B. 1. prs. 13, 14, 15.)

Complete .

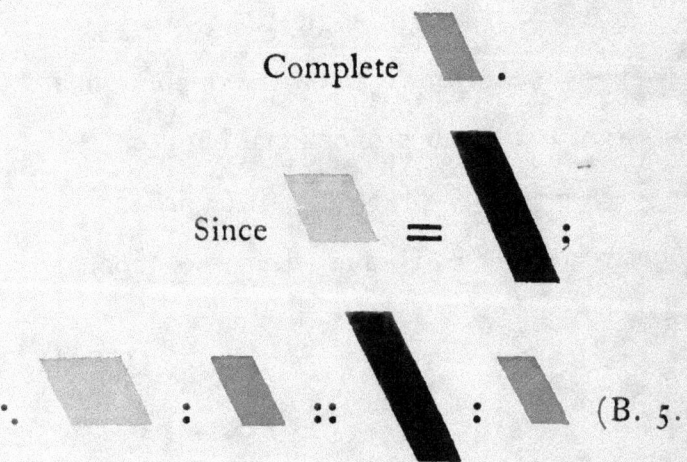

Since = ;

∴ : :: : (B. 5. pr. 7.)

∴ ━━━ : ━━━ :: ━━━ : ━━━

(B. 6. pr. 1.)

The fame conftruction remaining :

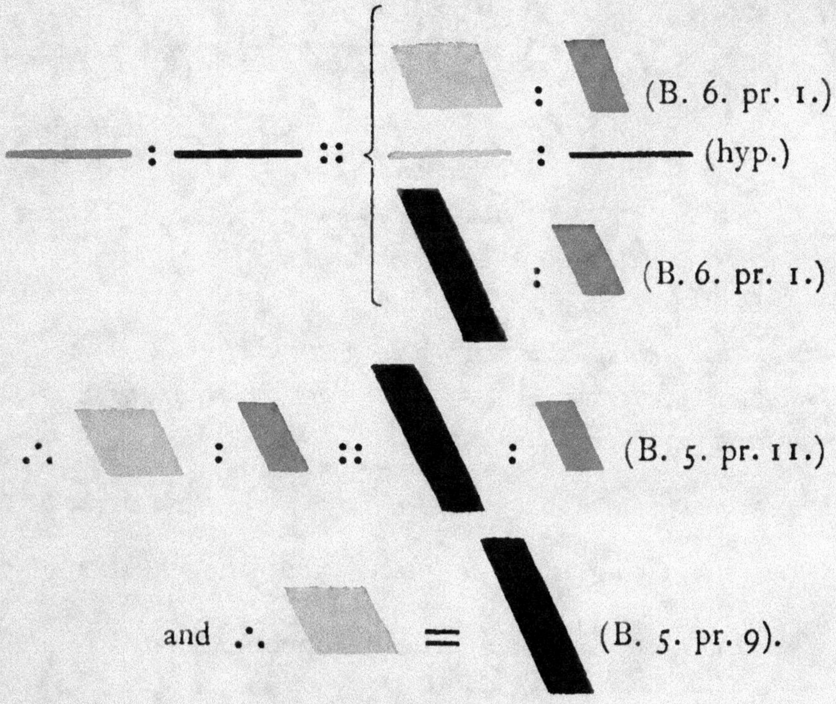

: (B. 6. pr. 1.)

━━━ : ━━━ :: { : (hyp.)

: (B. 6. pr. 1.)

∴ : :: : (B. 5. pr. 11.)

and ∴ = (B. 5. pr. 9).

Q. E. D.

H H

233

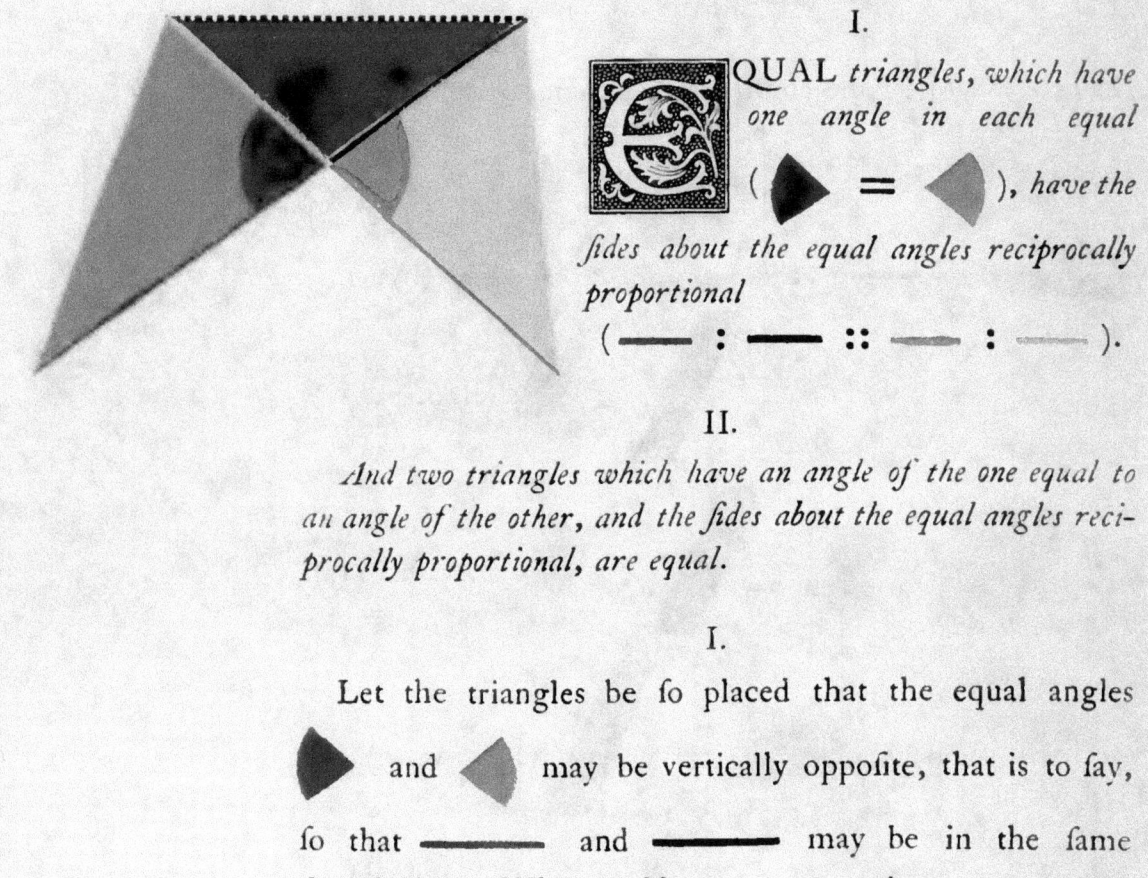

I.

E QUAL *triangles, which have one angle in each equal*
(◢ = ◣), *have the sides about the equal angles reciprocally proportional*
(—— : —— :: —— : ——).

II.

And two triangles which have an angle of the one equal to an angle of the other, and the sides about the equal angles reciprocally proportional, are equal.

I.

Let the triangles be fo placed that the equal angles ◢ and ◣ may be vertically oppofite, that is to fay, fo that —— and —— may be in the fame ftraight line. Whence alfo —— and —— muft be in the fame ftraight line. (B. 1. pr. 14.)

Draw ┄┄┄┄┄ , then

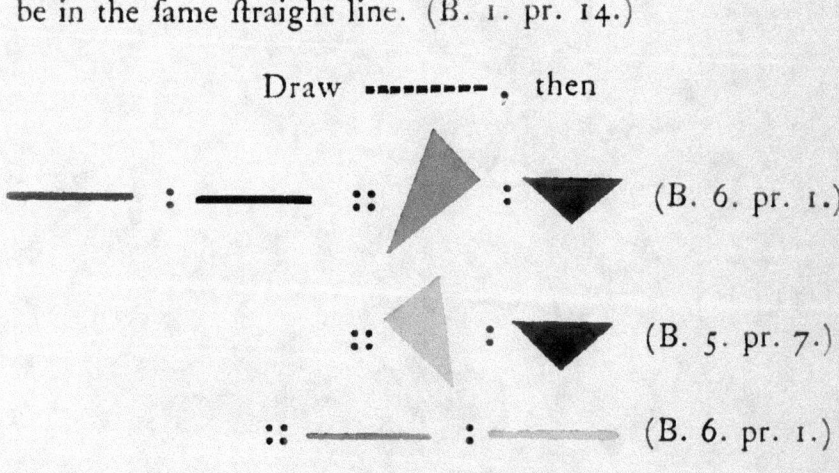

—— : —— :: ◣ : ▼ (B. 6. pr. 1.)

:: ◣ : ▼ (B. 5. pr. 7.)

:: —— : —— (B. 6. pr. 1.)

∴ ▬▬ : ▬▬ :: ▬▬ : ▬▬

(B. 5. pr. 11.)

II.

Let the same conſtruction remain, and

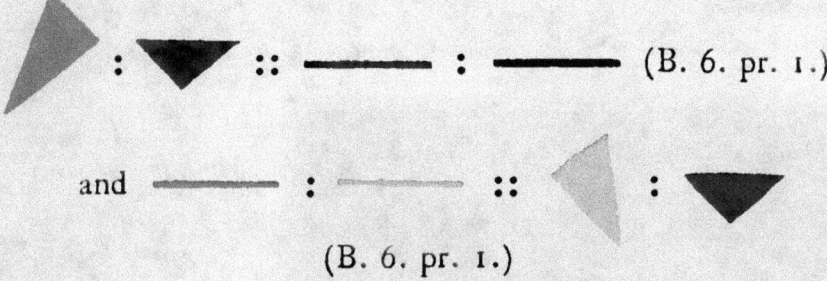

(B. 6. pr. 1.)

and ▬▬ : ▬▬ :: ◢ : ▼ (B. 6. pr. 1.)

(B. 6. pr. 1.)

But ▬▬ : ▬▬ :: ▬▬ : ▬▬ , (hyp.)

∴ ◢ : ▼ :: ◢ : ▼ (B. 5. pr. 11);

∴ ▶ = ◀ (B. 5. pr. 9.)

Q. E. D.

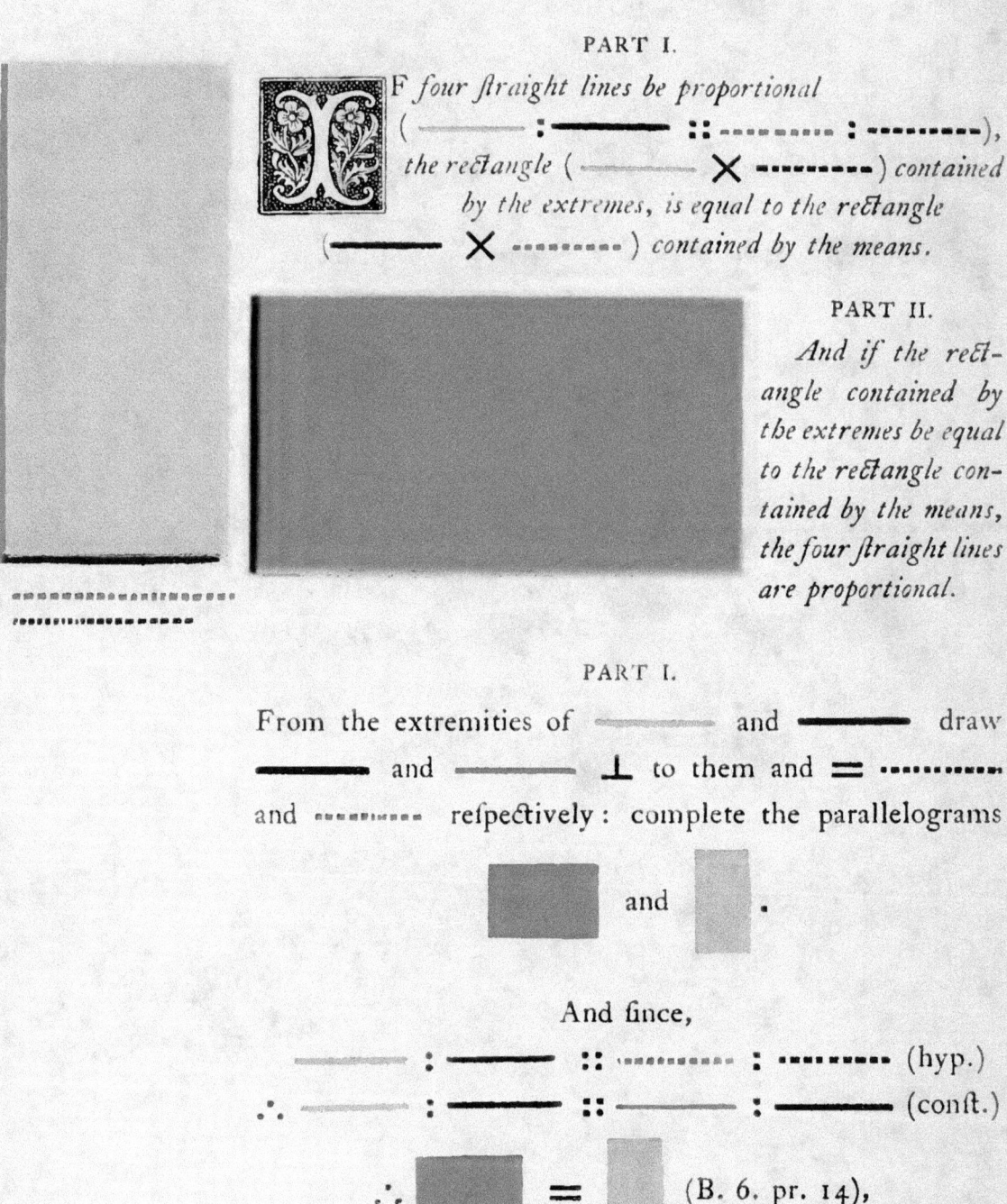

PART I.

IF *four ſtraight lines be proportional*
(—— : —— :: —— : ——),
the reƈtangle (—— ✕ ——) *contained
by the extremes, is equal to the reƈtangle*
(—— ✕ ——) *contained by the means.*

PART II.

*And if the reƈt-
angle contained by
the extremes be equal
to the reƈtangle con-
tained by the means,
the four ſtraight lines
are proportional.*

PART I.

From the extremities of —— and —— draw
—— and —— ⊥ to them and = ——
and —— reſpeƈtively: complete the parallelograms
and .

And ſince,

—— : —— :: —— : —— (hyp.)

∴ —— : —— :: —— : —— (conſt.)

∴ = (B. 6. pr. 14),

236

that is, the rectangle contained by the extremes, equal to
the rectangle contained by the means.

PART II.

Let the fame conftruction remain; becaufe

$$\text{------} = \text{------} , \quad \blacksquare = \blacksquare$$

$$\text{and} \quad \text{------} = \text{------} .$$

$$\therefore \quad \text{------} : \text{------} :: \text{------} : \text{------}$$

(B. 6. pr. 14).

$$\text{But} \quad \text{------} = \text{------} ,$$

$$\text{and} \quad \text{------} = \text{------} \text{ (conft.)}$$

$$\therefore \quad \text{------} : \text{------} :: \text{------} : \text{------}$$

(B. 5. pr. 7).

Q. E. D.

PART I

F *three ftraight lines be pro-portional (* ——— : ——— :: ——— : ——— *) the rectangle under the extremes is equal to the fquare of the mean.*

PART II.

And if the rectangle under the extremes be equal to the fquare of the mean, the three ftraight lines are proportional.

PART I.

Affume ——— = ———, and

fince ——— : ——— :: ——— : ———,

then ——— : ——— :: ——— : ———,

∴ ——— × ——— = ——— × ———

(B. 6. pr. 16).

But ——— = ———,

∴ ——— × ——— = ——— × ———,

or = ———^{2}; therefore, if the three ftraight lines are proportional, the rectangle contained by the extremes is equal to the fquare of the mean.

PART II.

Affume ——— = ———, then

——— × ——— = ——— × ———,

∴ ——— : ——— :: ——— : ———

(B. 6. pr. 16), and

∴ ——— : ——— :: ——— : ———.

Q. E. D.

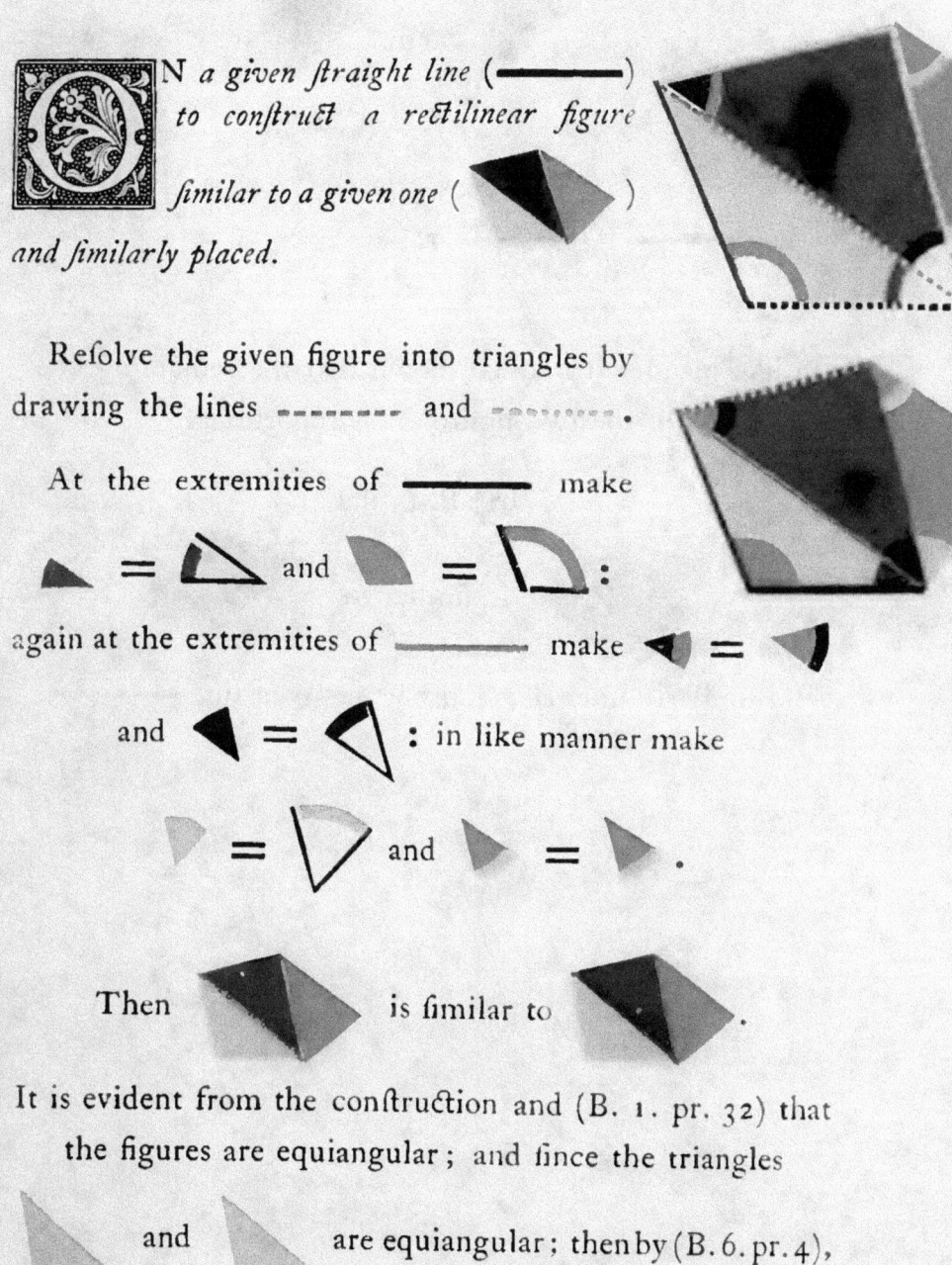

O̲N a given ſtraight line (━━━) to conſtruct a rectilinear figure ſimilar to a given one ( ) and ſimilarly placed.

Reſolve the given figure into triangles by drawing the lines ┈┈┈┈ and ┉┉┉.

At the extremities of ━━━ make

▰ = ◹ and ◗ = ◠ :

again at the extremities of ━━━ make ◀ = ◣

and ◀ = ◁ : in like manner make

◣ = ◺ and ▶ = ◣ .

Then is ſimilar to .

It is evident from the conſtruction and (B. 1. pr. 32) that the figures are equiangular ; and ſince the triangles

and are equiangular ; then by (B. 6. pr. 4),

━━ : ━━ :: ┉┉ : ━━

and ━━ : ━━ :: ━━ : ┈┈ .

Again, becauſe 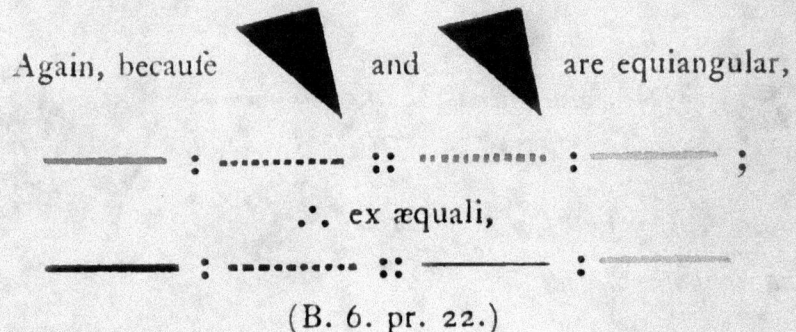 and are equiangular,

▬▬▬ : ‥‥‥‥ :: ‥‥‥‥ : ▬▬▬ ;

∴ ex æquali,

▬▬▬ : ‥‥‥ :: ▬▬▬ : ▬▬▬

(B. 6. pr. 22.)

In like manner it may be ſhown that the remaining ſides of the two figures are proportional.

∴ by (B. 6. def. 1.)

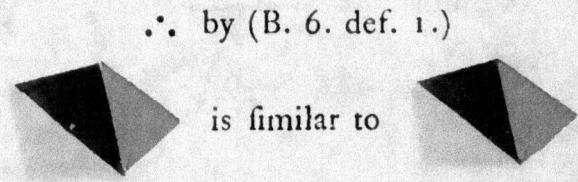 is ſimilar to

and ſimilarly ſituated ; and on the given line ▬▬▬ .

Q. E. D.

IMILAR *trian-*

gles (

and) *are to one another in the duplicate ratio of their homologous fides.*

Let and be equal angles, and ▪▪▪▪▪▪ ▬▬▬▬

and ▬▬▬▬▬ homologous fides of the fimilar triangles

and and on ▪▪▪▪▪▪ ▬▬▬▬ the greater

of thefe lines take ▪▪▪▪▪▪ a third proportional, fo that

▪▪▪▪▪ ▬▬▬ : ▬▬▬ :: ▬▬▬ : ▪▪▪▪▪▪▪▪ ;

draw ▪▪▪▪▪▪▪▪ .

▪▪▪▪▪ ▬▬▬ : ▬▬▬▬ :: ▬▬▬ : ▬▬▬▬

(B. 6. pr. 4) ;

∴ ▪▪▪▪▪ ▬▬▬ : ▬▬▬ :: ▬▬▬ : ▬▬▬

(B. 5. pr. 16, alt.),

but ▪▪▪▪▪ : ▬▬▬ :: ▬▬▬ : ▪▪▪▪▪ (conſt.),

∴ ▬▬▬ : ▪▪▪▪▪ :: ▬▬▬ : ▬▬▬ confe-

I I

quently ▲ = ◣ for they have the ſides about

the equal angles ◣ and ▲ reciprocally proportional
(B. 6. pr. 15);

∴ ◣ : ▲ :: ◣ : ◣
(B. 5 pr. 7);

but ◣ : ◣ :: ▬▬▬ : ▬▬▬
(B. 6. pr. 1),

∴ ◣ : ▲ :: ▬▬▬ : ▬▬▬ ,

that is to ſay, the triangles are to one another in the duplicate ratio of their homologous ſides
▬▬▬ and ▬▬▬ (B. 5. def. 11).

Q. E. D.

IMILAR *poly-gons may be di-vided into the fame number of fimilar triangles, each fimilar pair of which are propor-tional to the polygons; and the polygons are to each other in the duplicate ratio of their homologous fides.*

Draw ———— and ▪▪▪▪▪▪▪▪, and ———— and ----------, refolving the polygons into triangles. Then becaufe the polygons are fimilar, ◣ = ◣,

and ———— : ▪▪▪▪▪▪ :: ———— : ----------

∴ ◿ and ◿ are fimilar, and ◀ = ◀ (B. 6. pr. 6);

but ◆ = ◆ becaufe they are angles of fimilar poly-gons; therefore the remainders ▲ and ▲ are equal; hence ▪▪▪▪▪▪▪ : ---------- :: ---------- : ▪▪▪▪▪▪▪, on account of the fimilar triangles, and ⸱⸱⸱⸱⸱⸱⸱⸱ : —— :: ⸱⸱⸱⸱⸱⸱⸱ : —— ,

on account of the fimilar polygons,

∴ ⸱⸱⸱⸱⸱⸱⸱ : —— :: ⸱⸱⸱⸱⸱⸱⸱ : —— ,

ex æquali (B. 5. pr. 22), and as thefe proportional fides

contain equal angles, the triangles 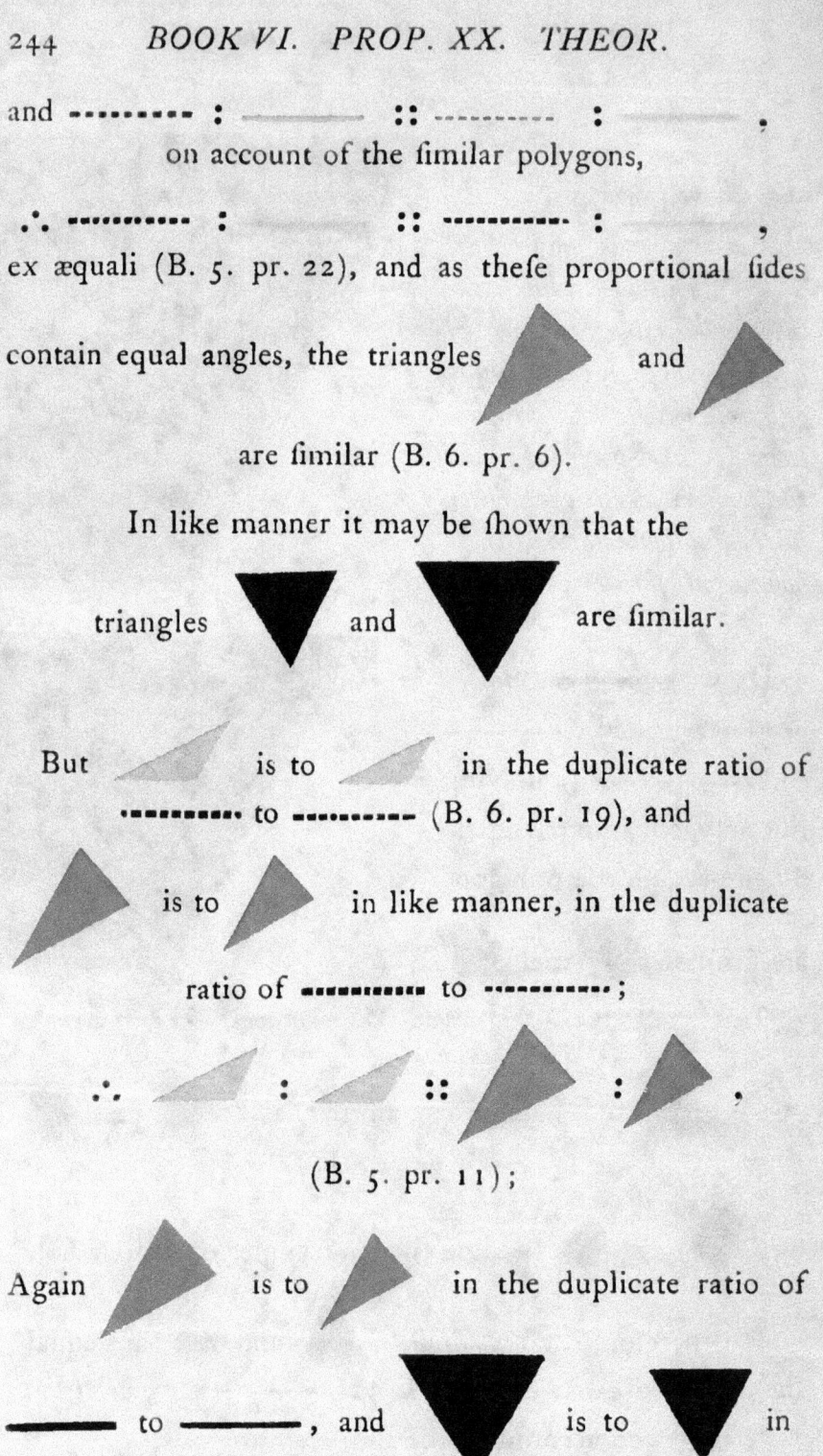 and

are fimilar (B. 6. pr. 6).

In like manner it may be fhown that the

triangles and are fimilar.

But is to in the duplicate ratio of

⸱⸱⸱⸱⸱⸱⸱ to ⸱⸱⸱⸱⸱⸱⸱ (B. 6. pr. 19), and

is to in like manner, in the duplicate

ratio of ⸱⸱⸱⸱⸱⸱⸱ to ⸱⸱⸱⸱⸱⸱⸱ ;

∴ : :: : ,

(B. 5. pr. 11);

Again is to in the duplicate ratio of

—— to —— , and is to in

the duplicate ratio of ———— to ————.

and as one of the antecedents is to one of the confequents, fo is the fum of all the antecedents to the fum of all the confequents; that is to fay, the fimilar triangles have to one another the fame ratio as the polygons (B. 5. pr. 12).

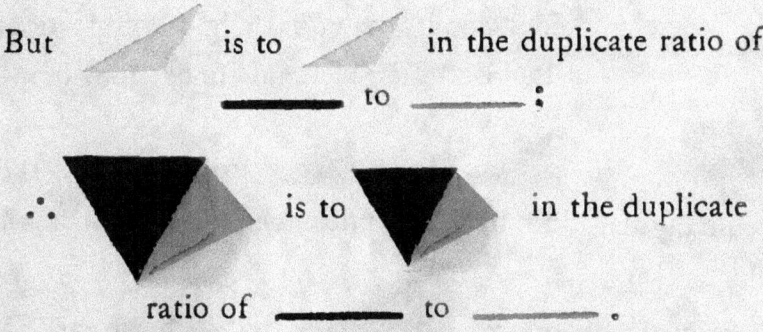

Q. E. D

ECTILINEAR *figures*

(*and*)

which are similar to the same figure ( )
are similar also to each other.

Since and are simi-
lar, they are equiangular, and have the
fides about the equal angles proportional
(B. 6. def. 1); and fince the figures

and are alfo fimilar, they
are equiangular, and have the fides about the equal angles

proportional; therefore and are alfo
equiangular, and have the fides about the equal angles proportional (B. 5. pr. 11), and are therefore fimilar.

Q. E. D.

PART I.

F *four straight lines be pro-portional* (———— : ———— :: ——— : ————), *the similar rectilinear figures similarly described on them are also proportional.*

PART II.

And if four similar rectilinear figures, similarly described on four straight lines, be proportional, the straight lines are also proportional.

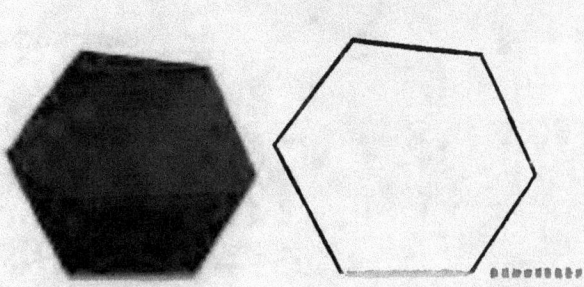

PART I.

Take ---------- a third proportional to ————

and ————, and ••••••••••• a third proportional

to ——— and ——— (B. 6. pr. 11);

since ———— : ———— :: ——— : ——— (hyp.),

———— : •••••••••• :: ——— : •••••••••• (conft.)

∴ ex æquali,

———— : ---------- :: ——— : ••••••••••• ;

but ▲ : ▲ :: ———— : ••••••••••

(B. 6. pr. 20),

and ⬡ : ⬡ :: ——— : •••••••••• ;

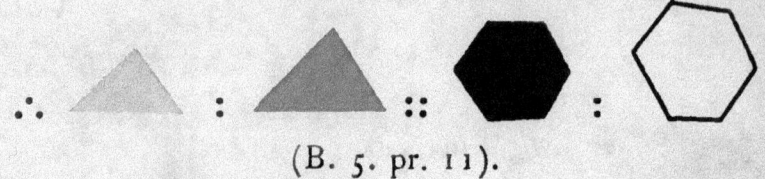

∴   ◢ : ◢ :: ⬡ : ⬡

(B. 5. pr. 11).

PART II.

Let the fame conftruction remain:

◢ : ◢ :: ⬡ : ⬡   (hyp.),

∴ ▬▬▬ : ••••••••• :: ▬▬▬ : ------- (conft.)

and ∴ ▬▬▬ : ▬▬▬ :: ▬▬▬ : ▬▬▬ .

(B. 5. pr. 11).

Q. E. D.

QUIANGULAR *parallel-ograms* (and) *are to one another in a ratio compounded of the ratios of their sides.*

Let two of the sides ———— and .... about the equal angles be placed so that they may form one straight line.

Since ▼ + ◗ = ◲ ,

and ◣ = ▼ (hyp.),

◣ + ◗ = ◲ ,

and ∴ ———— and ———— form one straight line (B. 1. pr. 14);

complete ▱ .

Since ▱ : ▱ :: ———— : ———— (B. 6. pr. 1),

and ▱ : ▰ :: ———— : ———— (B. 6. pr. 1),

▱ has to ▰ a ratio compounded of the ratios of ———— to ...., and of ———— to ———— .

ᴋ ᴋ Q. E. D.

I N any parallelogram ( ) the parallelograms (and) which are about the diagonal are similar to the whole, and to each other.

As and have a common angle they are equiangular; but becaufe ——— || ———

▲ and ▲▲ are similar (B. 6. pr. 4),

∴ ——— : ——— :: ——— : ——— ;

and the remaining oppofite fides are equal to thofe,

∴ and have the fides about the equal angles proportional, and are therefore fimilar.

In the fame manner it can be demonftrated that the

parallelograms and are fimilar.

Since, therefore, each of the parallelograms

 and is fimilar to , they are fimilar to each other.

Q. E. D.

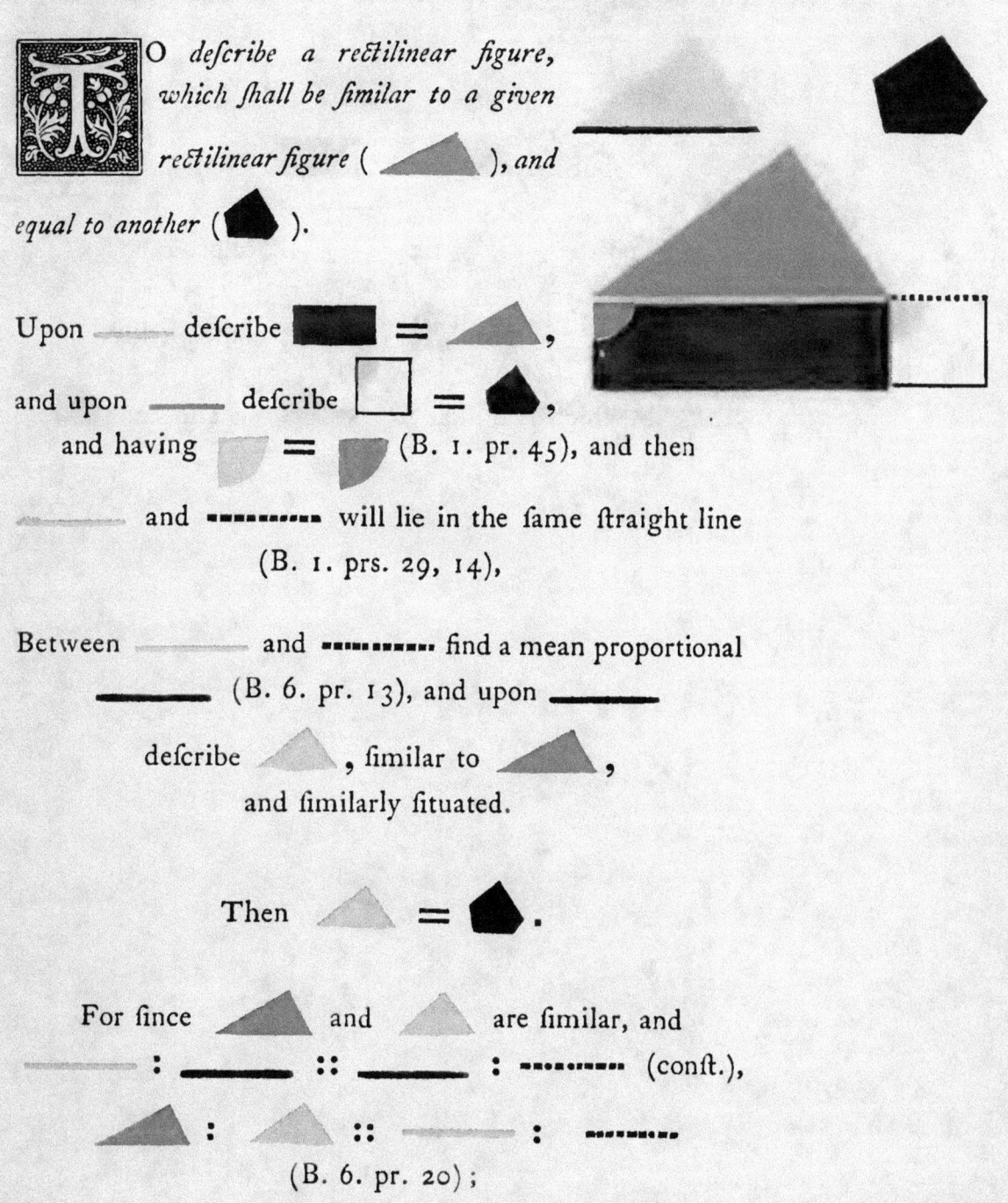

O *defcribe a rectilinear figure, which fhall be fimilar to a given rectilinear figure (   ), and equal to another (   ).*

Upon ____ defcribe ▬ = ▲,

and upon ____ defcribe □ = ◆,

and having ◢ = ◣ (B. 1. pr. 45), and then

____ and ••••••••• will lie in the fame ftraight line (B. 1. prs. 29, 14),

Between ____ and ••••••••• find a mean proportional ____ (B. 6. pr. 13), and upon ____

defcribe ▲, fimilar to ▲, and fimilarly fituated.

Then ▲ = ◆.

For fince ▲ and ▲ are fimilar, and

____ : ▬ :: ____ : ••••••••• (conft.),

▲ : ▲ :: ____ : ••••••••• (B. 6. pr. 20);

251

but ▬ : ▢ :: ———— : ------- (B. 6. pr. 1);

∴ ◣ : ◺ :: ▬ : ▢ (B. 5. pr. 11);

but ◣ = ▬ (conſt.),

and ∴ = ▢ (B. 5. pr. 14);

and ▢ = ⬟ (conſt.); conſequently,

which is ſimilar to ◣ is alſo = ⬟.

Q. E. D.

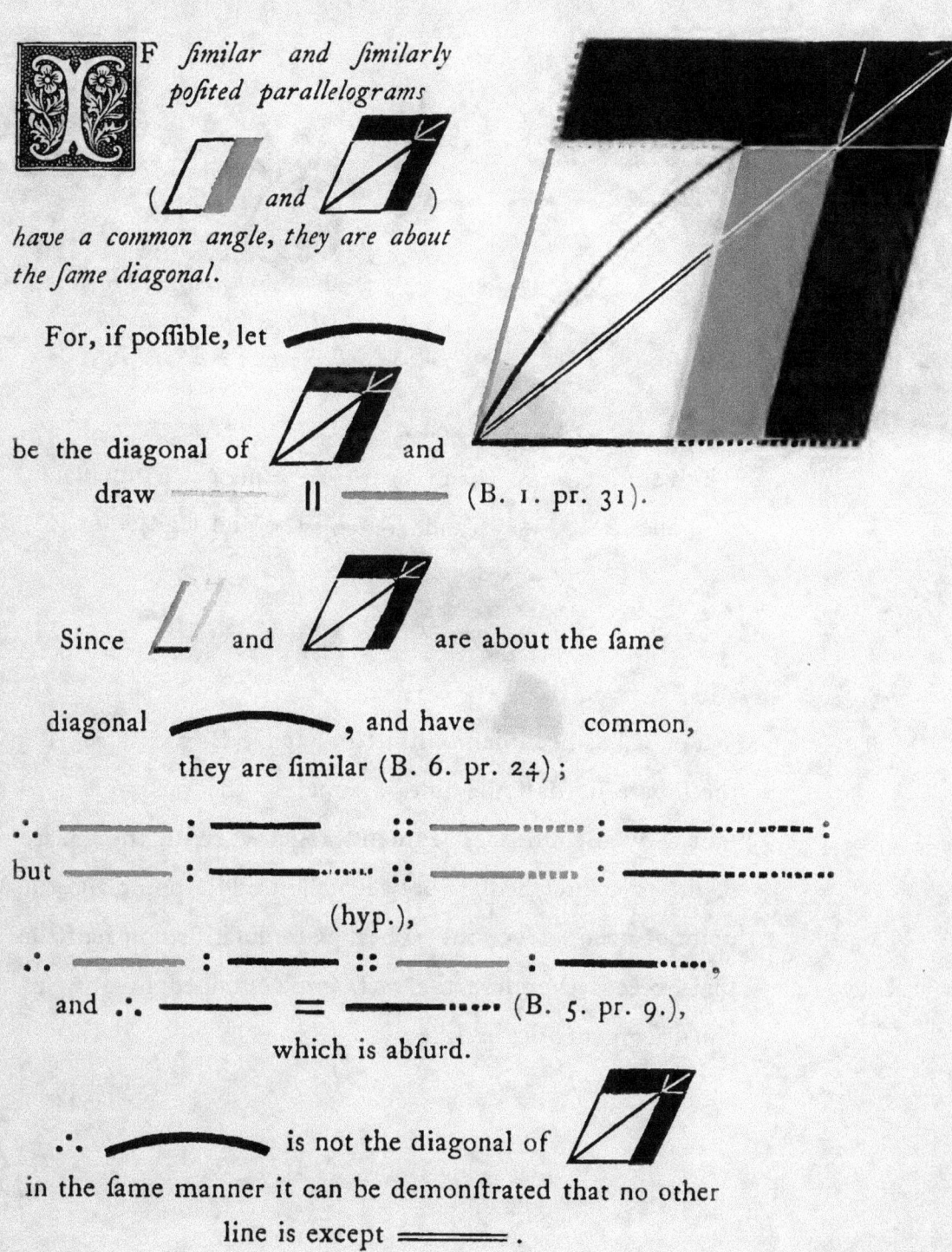

F *similar and similarly posited parallelograms*

(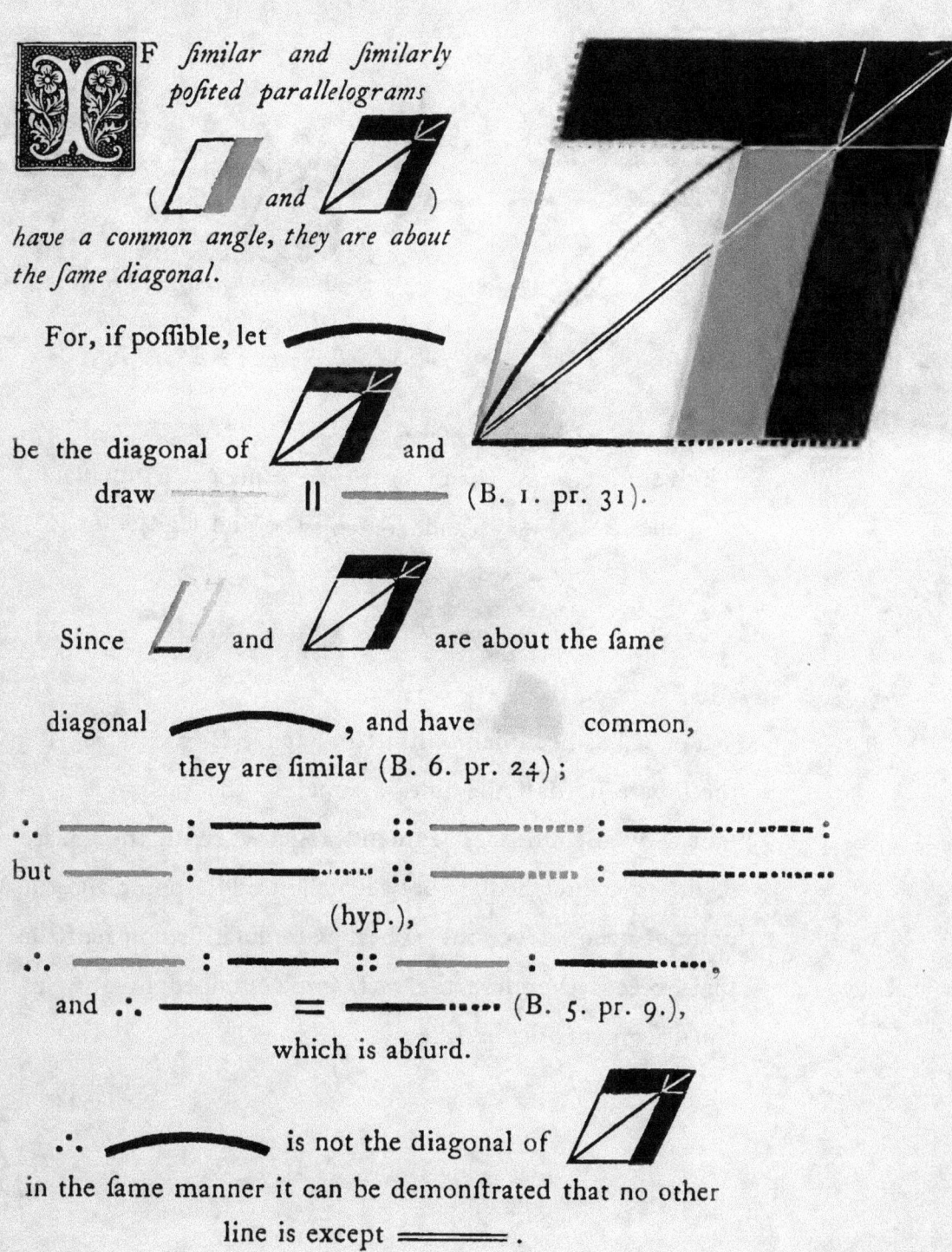 *and* 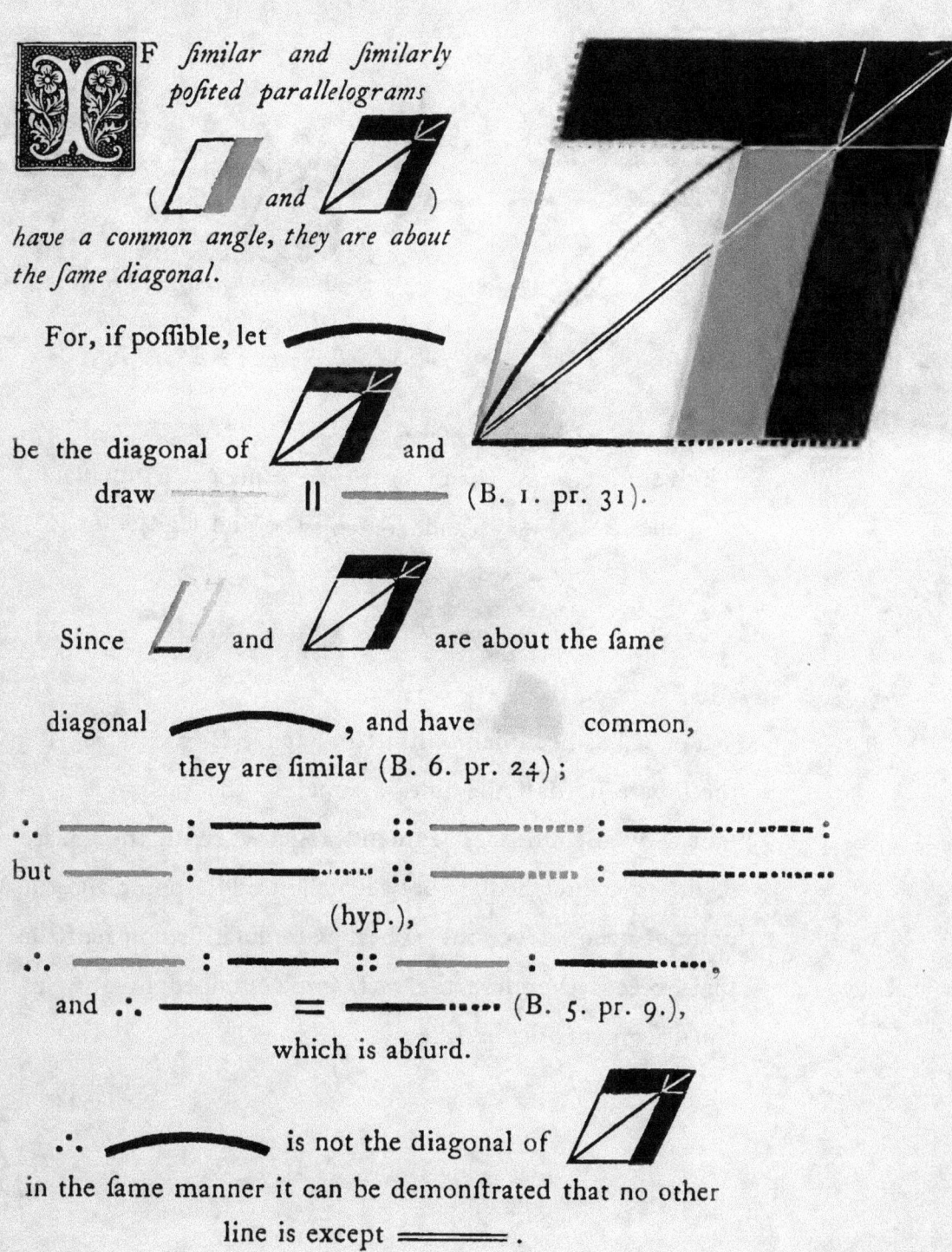)

have a common angle, they are about the same diagonal.

For, if possible, let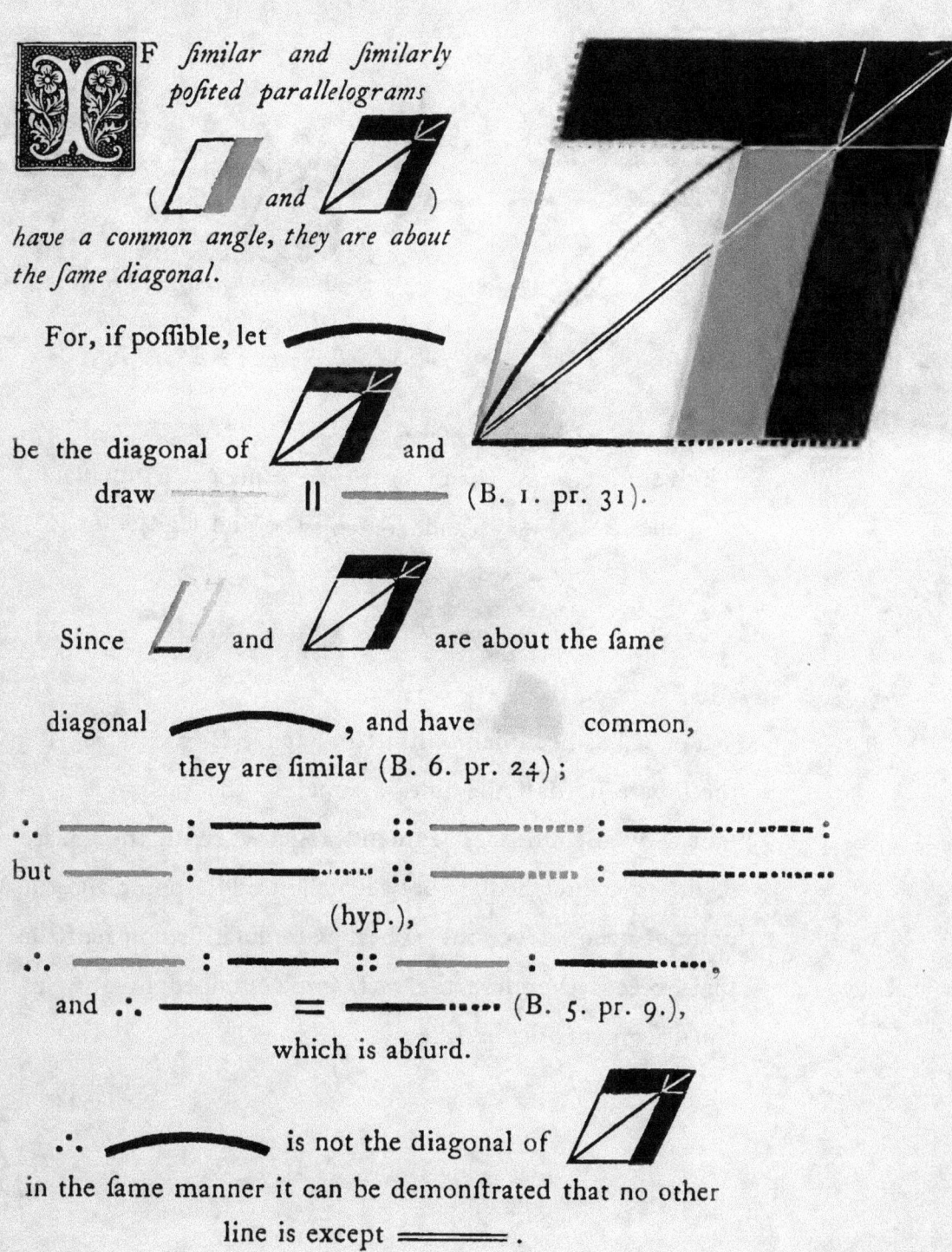

be the diagonal of and

draw ‖ (B. 1. pr. 31).

Since and are about the same

diagonal , and have common,

they are similar (B. 6. pr. 24);

∴ : :: :

but : :: :

(hyp.),

∴ : :: : ,

and ∴ = (B. 5. pr. 9.),

which is absurd.

∴ is not the diagonal of

in the same manner it can be demonstrated that no other

line is except .

Q. E. D.

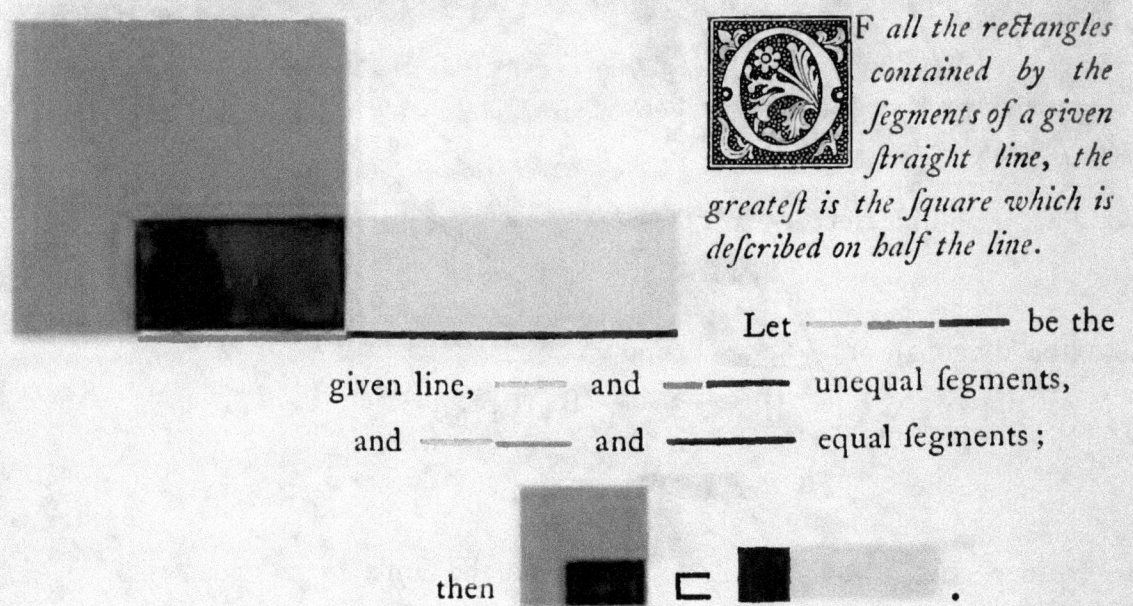

F all the rectangles contained by the segments of a given straight line, the greatest is the square which is described on half the line.

Let ▬▬▬ be the given line, ▬▬ and ▬▬ unequal segments, and ▬▬ and ▬▬ equal segments;

then ■ ⊏ ■ .

For it has been demonſtrated already (B. 2. pr. 5), that the ſquare of half the line is equal to the rectangle contained by any unequal ſegments together with the ſquare of the part intermediate between the middle point and the point of unequal ſection. The ſquare deſcribed on half the line exceeds therefore the rectangle contained by any unequal ſegments of the line.

Q. E. D.

 O *divide a given straight line* (━━━━━━) *so that the rec-tangle contained by its segments may be equal to a given area, not exceeding the square of half the line.*

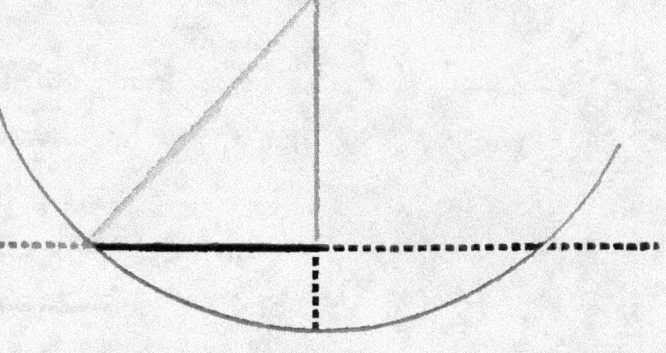

Let the given area be ═ ━━━━━².

Bifect ━━━━━, or

make ━━━━━ ═ ━━━━;

and if ━━━━² ═ ━━━━²,

the problem is folved.

But if ━━━━² ╪ ━━━━², then

muft ━━━━ ⊏ ━━━━ (hyp.).

Draw ━━━━ ⊥ ━━━━ ═ ━━━━;

make ━━━━ ═ ━━━━ or ━━━━;

with ━━━━ as radius defcribe a circle cutting the given line; draw ━━━━.

Then ━━━ ✕ ━━━━ ╋ ━━━━² ═ ━━━━²

(B. 2. pr. 5.) ═ ━━━━².

But ━━━━² ═ ━━━━² ╋ ━━━━²

(B. 1. pr. 47);

$$\therefore \ \text{------} \ \times \ \text{------} \ + \ \text{------}^2$$

$$= \ \text{------}^2 \ + \ \text{------}^2,$$

from both, take ——————^{2},

and ------ $\times$ ———————— $=$ ——————^{2}.

But ————— $=$ ———— (conſt.),

and $\therefore$ —————————— is ſo divided

that ------ $\times$ ———————— $=$ ————^{2}.

Q. E. D.

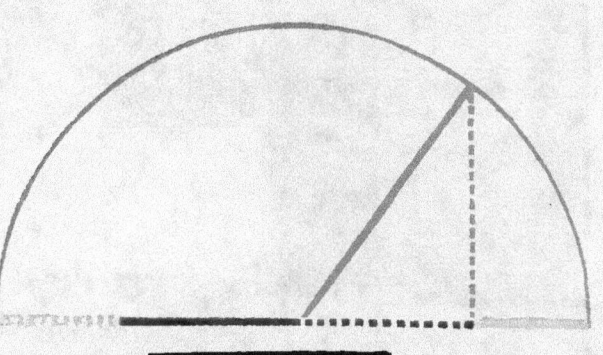

O *produce a given straight line* (———— ·······), *so that the rectangle contained by the segments between the extremities of the given line and the point to which it is produced, may be equal to a given area, i. e. equal to the square on* ——— .

Make ———— = ········· , and

draw ········· ⊥ ········· = ———— ;

draw ————— ; and

with the radius ————— , defcribe a circle

meeting ———— ······· produced.

Then ———— ······· ———— × ———— + ·········^{2} =

·······^{2} (B. 2. pr. 6.) = —————^{2}.

But ————^{2} = ·········^{2} + ·········^{2} (B. 1. pr. 47.)

∴ ———— ···· ———— × ———— + ·········^{2} =

·········^{2} + ·········^{2},

from both take ·········^{2},

and ———— ······· ———— × ———— = ·········^{2} ;

but ········· = ———— ,

∴ ·········^{2} = the given area.

Q. E. D.

L L

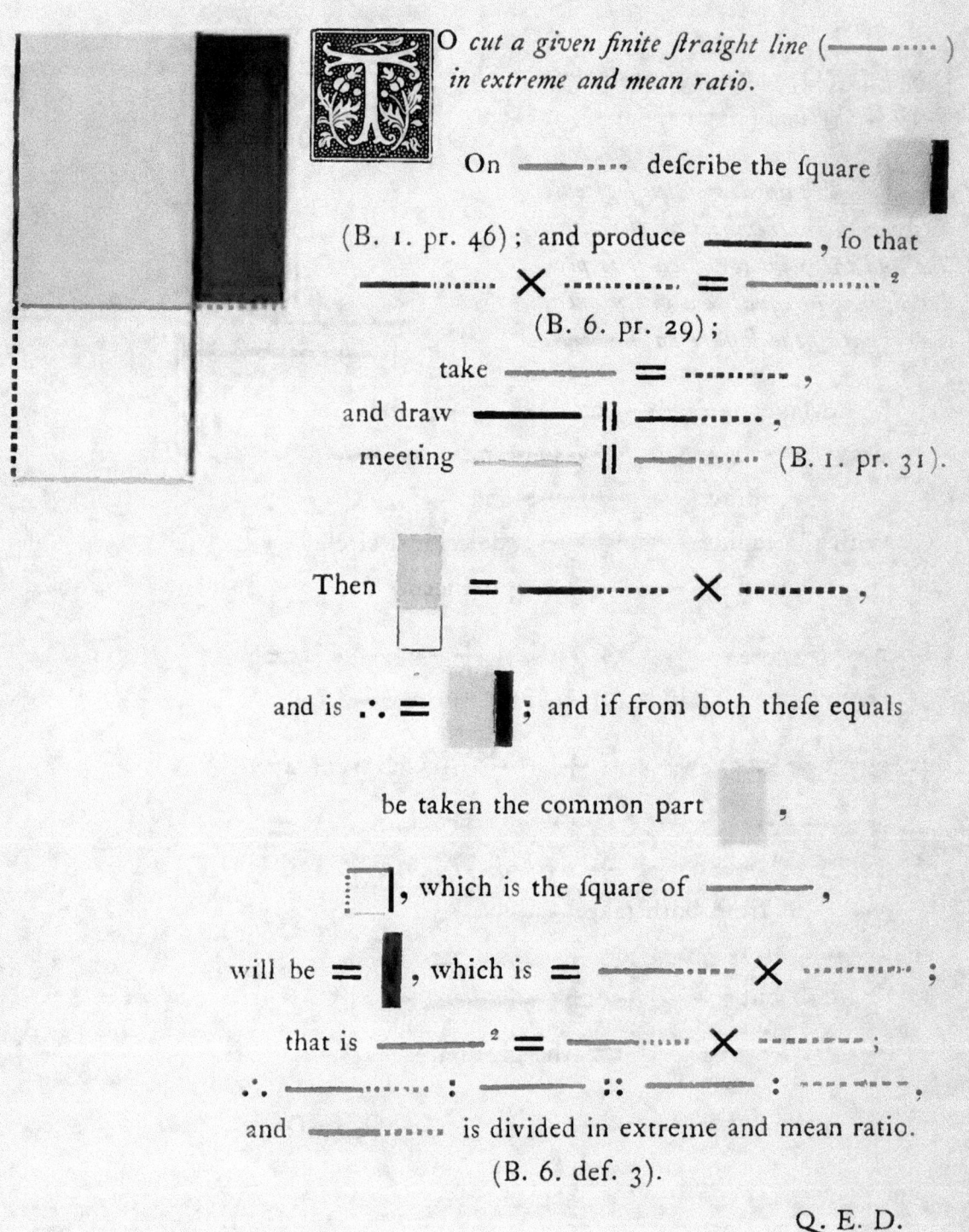

To cut a given finite straight line (——— ······) in extreme and mean ratio.

On ——— ······ describe the square (B. 1. pr. 46); and produce ———, so that

——— ·········· ✕ ·············· = ——— ······²

(B. 6. pr. 29);

take ——— = ············· ,

and draw ——— ‖ ——— ······· ,

meeting ——— ‖ ——— ······· (B. 1. pr. 31).

Then ▯ = ——— ······· ✕ ············ ,

and is ∴ = ▮ ; and if from both these equals

be taken the common part ▯ ,

▯ , which is the square of ——— ,

will be = ▮ , which is = ——— ······ ✕ ············ ;

that is ———² = ——— ······ ✕ ············ ;

∴ ——— ······ : ——— :: ——— : ··········· ,

and ——— ······ is divided in extreme and mean ratio.

(B. 6. def. 3).

Q. E. D.

F *any similar rectilinear figures be similarly described on the sides of a right an-*

gled triangle (————), the figure described on the side (————) sub-tending the right angle is equal to the sum of the figures on the other sides.

From the right angle draw ———— perpendicular

to ———— ;

then ———— : ———— :: ———— : ————

(B. 6. pr. 8).

∴ ▮ : ▮ :: ———— : ————

(B. 6. pr. 20).

but ▮ : ▮ :: ———— : ————

(B. 6. pr. 20).

Hence ———— + ———— : ————

:: ▮ + ▮ : ▮ ;

but ———— + ———— = ———— ;

and ∴ ▮ + ▮ = ▮ .

Q. E. D.

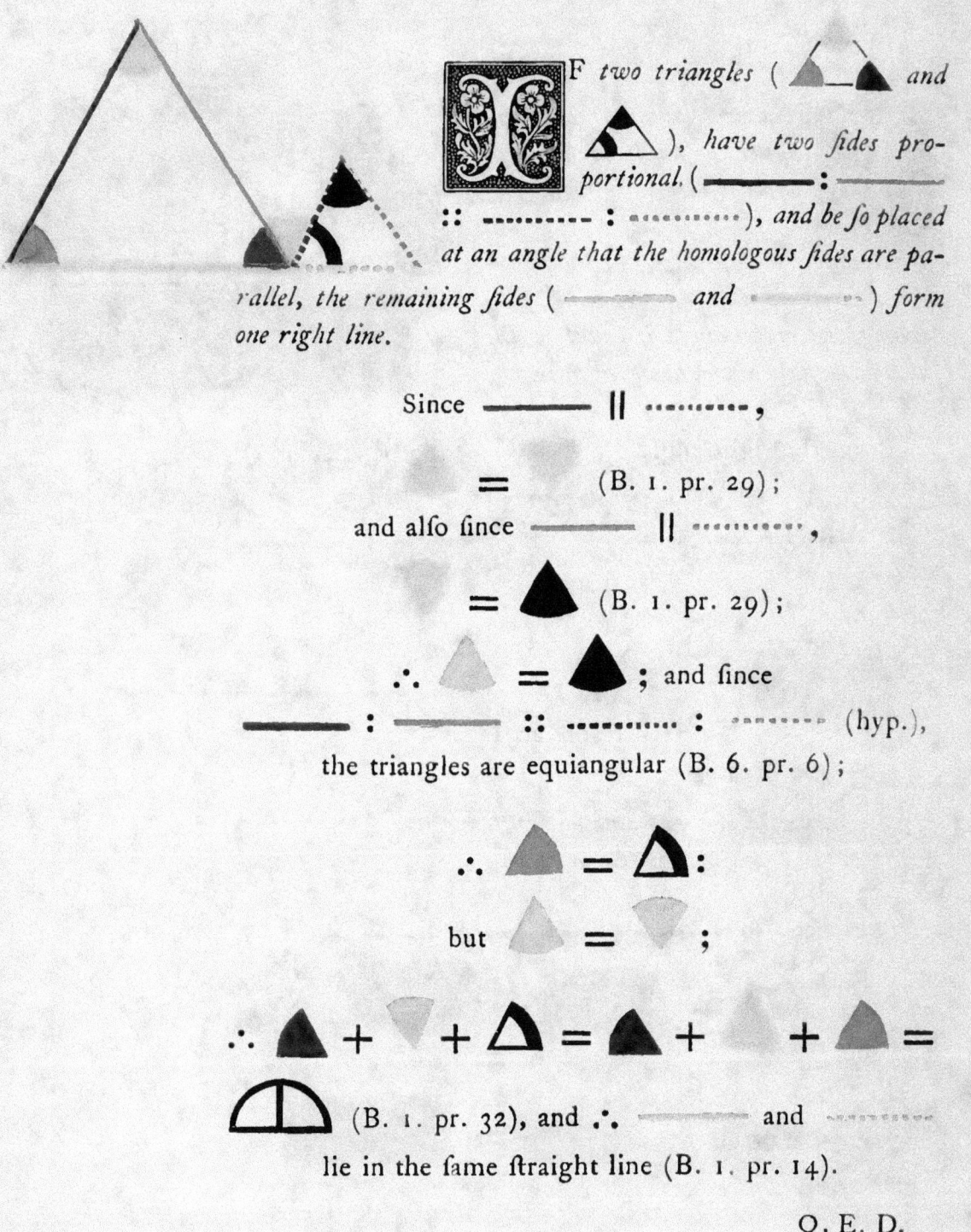

I F two triangles (and), have two fides proportional. (—— : —— :: .... : ··········), and be fo placed at an angle that the homologous fides are parallel, the remaining fides (——— and ———··) form one right line.

Since —— ‖ ········· ,

= (B. 1. pr. 29);

and alfo fince ——— ‖ ·········· ,

= (B. 1. pr. 29);

∴ = ; and fince

—— : —— :: ········· : ········· (hyp.),

the triangles are equiangular (B. 6. pr. 6);

∴ = :

but = ;

∴ + + = + + =

(B. 1. pr. 32), and ∴ ——— and ········· lie in the fame ftraight line (B. 1. pr. 14).

Q. E. D.

260

N equal circles (⬭ , ◯), angles, whether at the centre or circumference, are in the same ratio to one another as the arcs on which they ftand (▲ : ◢ :: ▬ : ⋯⋯); fo alfo are fectors.

Take in the circumference of ◯ any number of arcs ▬ , ▬ , &c. each ═ ▬ , and alfo in the circumference of ◯ take any number of arcs ⋯⋯ , ⋯⋯ , &c. each ═ ⋯⋯ , draw the radii to the extremities of the equal arcs.

Then fince the arcs ▬ , ▬ , ▬ , &c. are all equal, the angles ◢ , ◢ , ◣ , &c. are alfo equal (B. 3. pr. 27);

∴ ▲ is the fame multiple of ◢ which the arc ⌣ is of ▬ ; and in the fame manner ◣ is the fame multiple of ◢ , which the arc ⋯⋯⋯ is of the arc ⋯⋯ .

Then it is evident (B. 3. pr. 27),

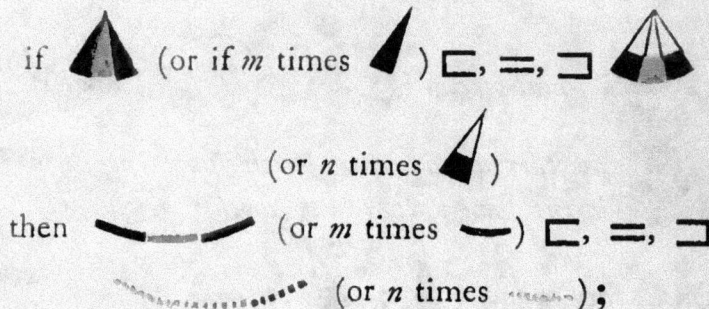

$$\therefore \quad \blacktriangle : \blacktriangle :: \; — : \; \text{......} \; , \text{(B. 5. def. 5), or the}$$

angles at the centre are as the arcs on which they ſtand; but the angles at the circumference being halves of the angles at the centre (B. 3. pr. 20) are in the ſame ratio (B. 5. pr. 15), and therefore are as the arcs on which they ſtand.

It is evident, that ſectors in equal circles, and on equal arcs are equal (B. 1. pr. 4; B. 3. prs. 24, 27, and def. 9). Hence, if the ſectors be ſubſtituted for the angles in the above demonſtration, the ſecond part of the propoſition will be eſtabliſhed, that is, in equal circles the ſectors have the ſame ratio to one another as the arcs on which they ſtand.

Q. E. D.

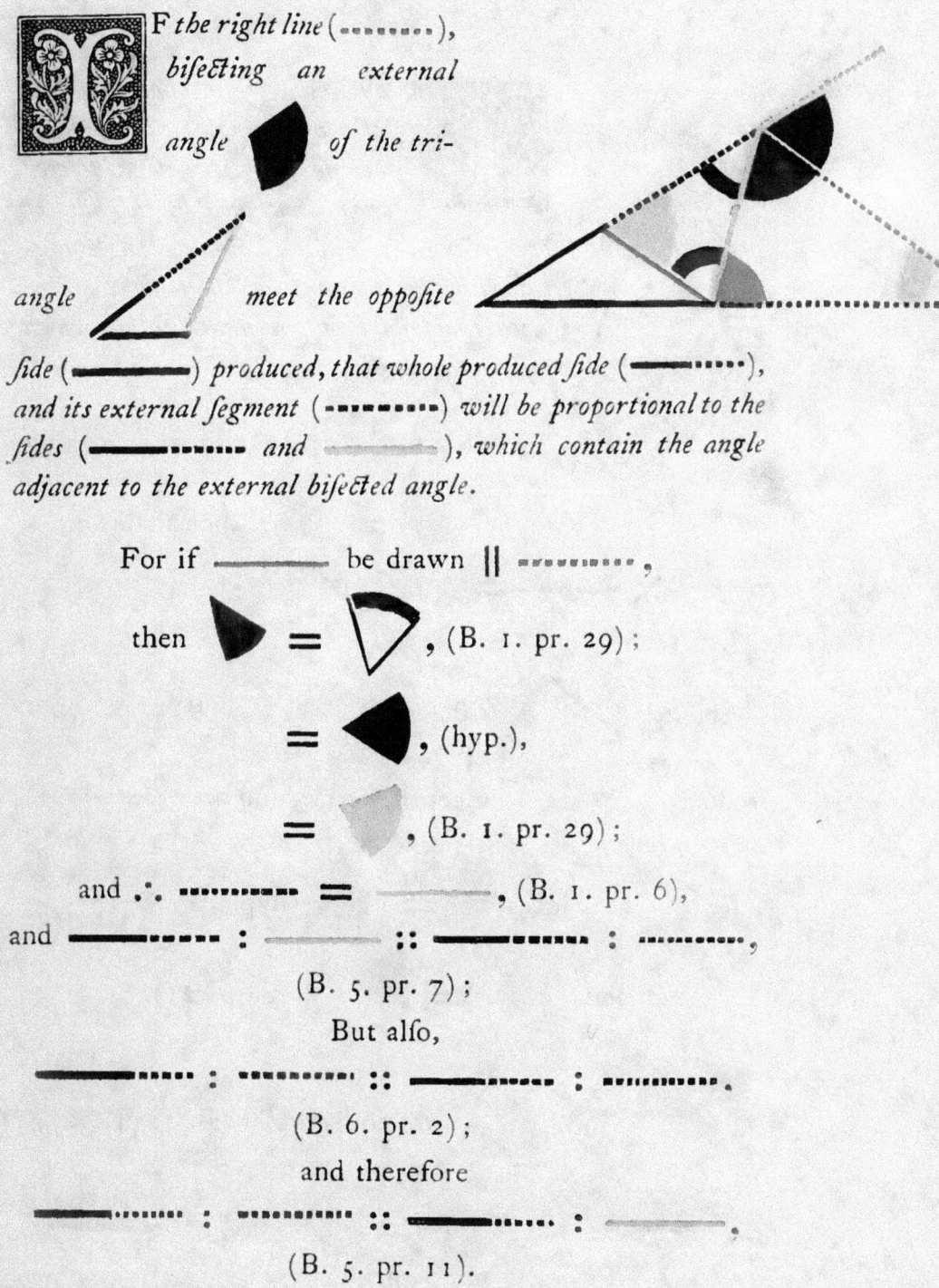

F *the right line* (-----------),
bisecting an external
angle *of the tri-*

angle *meet the opposite*

side (————) *produced, that whole produced side* (————----),
and its external segment (-----------) *will be proportional to the*
sides (————----- *and* ————), *which contain the angle*
adjacent to the external bisected angle.

For if ———— be drawn ‖ -----------,

then ◣ = ◺ , (B. 1. pr. 29);

= ◣ , (hyp.),

= ◿ , (B. 1. pr. 29);

and ∴ ----------- = ———— , (B. 1. pr. 6),

and ————----- : ———— :: ————---- : ----------- ,

(B. 5. pr. 7);

But also,

————----- : ----------- :: ————---- : ----------- ,

(B. 6. pr. 2);

and therefore

————----- : ----------- :: ————---- : ———— ,

(B. 5. pr. 11).

Q. E. D.

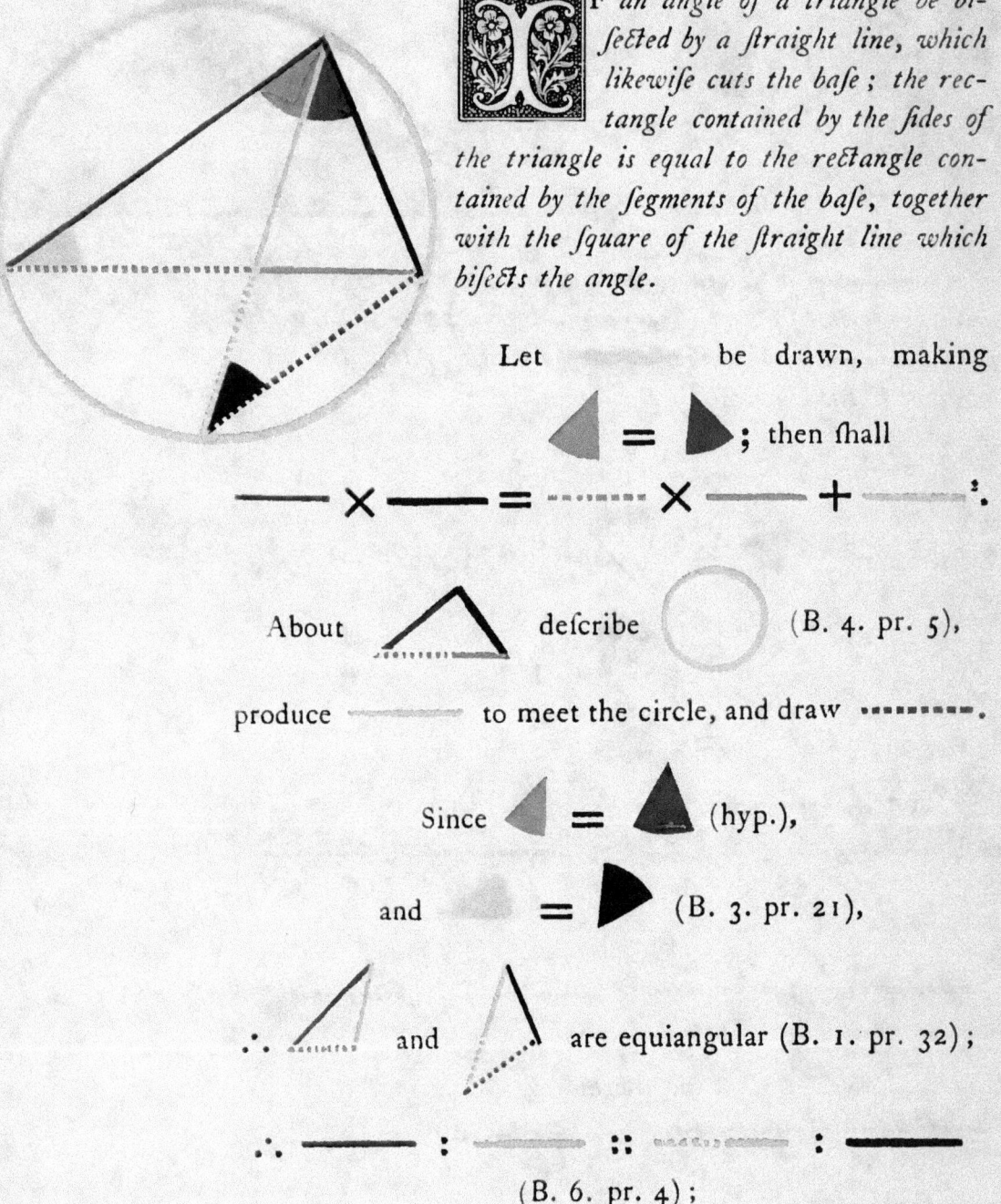

I F *an angle of a triangle be bi-*
sected by a straight line, which
likewise cuts the base; the rec-
tangle contained by the sides of
the triangle is equal to the rectangle con-
tained by the segments of the base, together
with the square of the straight line which
bisects the angle.

Let ———— be drawn, making

◣ = ◢ ; then shall

——— × —— = ······· × —— + ····· .

About △ describe ◯ (B. 4. pr. 5),

produce ———— to meet the circle, and draw ·······—.

Since ◢ = ▲ (hyp.),

and ◣ = ◢ (B. 3. pr. 21),

∴ ◺ and ◹ are equiangular (B. 1. pr. 32);

∴ —— : ······ :: ······ : ——

(B. 6. pr. 4);

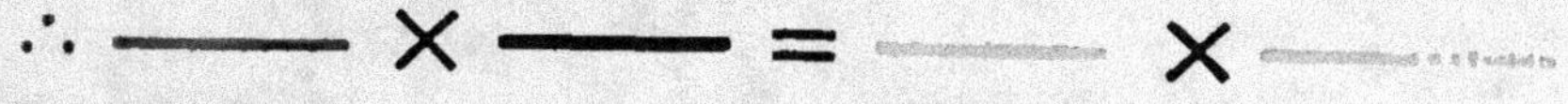

$\therefore$ ▬▬ × ▬▬ = ▬▬ × ▬▬

(B. 6. pr. 16.)

= ▬▬ × ▬▬ + ▬▬ ;

(B. 2. pr. 3);

but ▬▬ × ▬▬ = ▬▬ × ▬▬

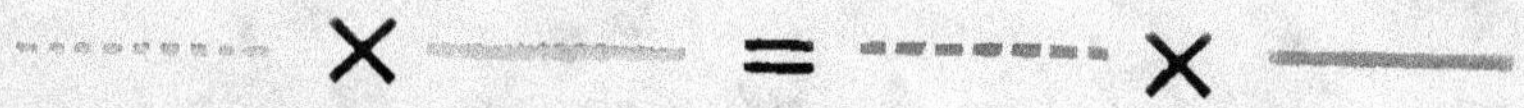

(B. 3. pr. 35);

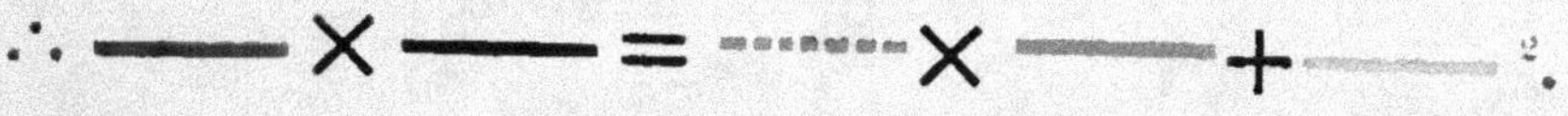

$\therefore$ ▬▬ × ▬▬ = ▬▬ × ▬▬ + ▬▬ .

Q. E. D.

M M

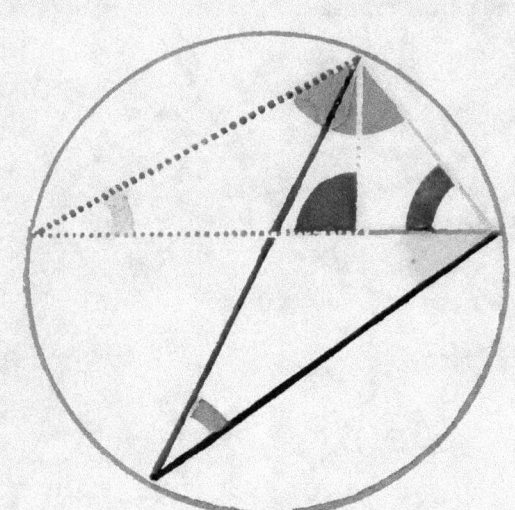

F *from any angle of a triangle a straight line be drawn perpendicular to the base; the rectangle contained by the sides of the triangle is equal to the rectangle contained by the perpendicular and the diameter of the circle described about the triangle.*

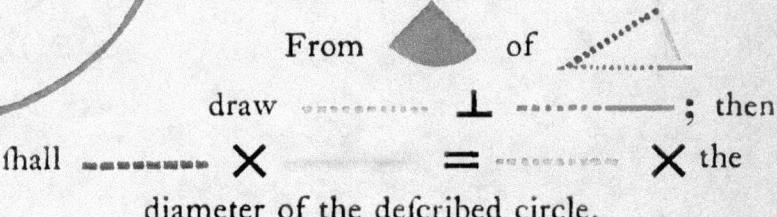

From ▱ of ◿

draw ⋯⋯ ⊥ ⋯⋯⋯ ; then

shall ▰ ✕ ▬ = ⋯⋯ ✕ the

diameter of the described circle.

Describe ◯ (B. 4. pr. 5), draw its diameter

▬▬▬ , and draw ▬▬▬ ; then because

▰ = ◠ (conft. and B. 3. pr. 31);

and ◁ = ◁ (B. 3. pr. 21);

∴ ◣ is equiangular to ◥ (B. 6. pr. 4);

∴ ▰▰ : ⋯⋯ :: ▬▬ : ▬▬ ;

and ∴ ▰▰ ✕ ▬▬ = ⋯⋯ ✕ ▬▬

(B. 6. pr. 16).

Q. E. D.

HE *rectangle contained by the diagonals of a quadrilateral figure inscribed in a circle, is equal to both the rectangles contained by* its opposite sides.

Let  be any quadrilateral

figure inscribed in ◯ ; and draw

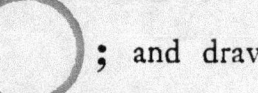 and ——— ; then

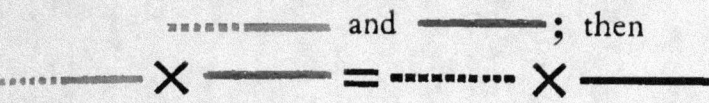

Make ▲ = ◣ (B. 1. pr. 23),

∴ ◢ = ◣ ; and ◤ = ◹ (B. 3. pr. 21);

∴ ——— : ——— :: ——— : ······ (B. 6. pr. 4);

and ∴ ——— × ——— = ——— × ····· (B. 6. pr. 16); again,

because ▲ = ◥ (conſt.),

267

and $\vee$ = $\diagdown$ (B. 3. pr. 21):

$\therefore$ ▬▬ : ▬▬ :: ▬▬ : ▬▬

(B. 6. pr. 4);

and $\therefore$ ▬▬ $\times$ ▬▬ = ▬▬ $\times$ ▬▬

(B. 6. pr. 16);

but, from above,

▬▬ $\times$ ▬▬ = ▬▬ $\times$ ▬▬ ;

$\therefore$ ▬▬ $\times$ ▬▬ = ▬▬ $\times$ ▬▬ + ▬▬ $\times$ ▬▬

(B. 2. pr. 1.

Q. E. D.

THE END.

CHISWICK: PRINTED BY C. WHITTINGHAM.

www.ingramcontent.com/pod-product-compliance
Lightning Source LLC
Chambersburg PA
CBHW080759180526
45168CB00006B/2270